EVOLUTIONARY WARS

EVOLUTIONARY WARS

A Three-Billion-Year Arms Race

The Battle of Species on Land, at Sea, and in the Air

CHARLES KINGSLEY LEVY

Illustrations by
Trudy Nicholson

W. H. Freeman and Company

New York

Text Design: Blake Logan

All illustrations, with the exception of those appearing on pages 5, 15, 50, 66, 68, 104, 125, 148, 170, 199, 212, 215, 224, 226, 239, 246, 249, 269, and 273 are reprinted with permission of Grass Instrument Division of Astro-Med, Inc.

Library of Congress Cataloging-in-Publication Data

Levy, Charles K., 1924–
    Evolutionary wars: a three-billion-year arms race: the battle of
species on land, at sea, and in the air/Charles Kingsley Levy;
illustrations by Trudy Nicholson.
        p.   cm.
    ISBN 0-7167-3483-4 (trade)
    1. Natural selection.   2. Predation (Biology)   I. Title.
    QH375.L54   1999
    576.8'2—dc21   99-39982
                                        CIP

Printed in the United States of America

First printing 1999

W. H. Freeman and Company
41 Madison Avenue, New York, NY 10010
Houndmills, Basingstoke RG21 6XS, England

To my wife, June

Contents

Preface

As a child growing up in the age before television, when parents read to children and grandparents told tales that demanded attentive listening, I was struck by the prevalence in their stories of a theme of conflict and struggle for survival. In religious school, the stories of David and Goliath, Samson and the Philistines, and the flight of the Hebrews from Egypt held my rapt attention. At home, my father, a Rudyard Kipling buff, read to me from *The Jungle Book* and the *Just So Stories*. I could visualize the man-child Mowgli, raised by wolves, thwarting the evil tiger, Sher Kahn. Another Kipling tale, about the brave little mongoose who protected his adopted human family from the malevolent cobras, Nag and Nagina, was also indelibly imprinted on my mind. It was the beginning of my lifelong fascination with life's struggles for survival, competition, and predator-prey interactions.

Like all children, I became fascinated with dinosaurs such as *Tyrannosaurus rex* and the monstrous herbivores they hunted. Walt Disney recognized the universality of our interest and exploited it in his early full-length cartoon films. Later, Steven Spielberg tapped into our fascination with evolutionary battles 100 million years old; his spectacular fantasy *Jurassic Park* included not only *T. rex* but also a pack of swift hunters called velociraptors.

I was fortunate to have a father who was a cartographer and explorer, and who spent many years in the jungles of Southeast Asia and along the Alaskan coast all the way to the tip of the Aleutians. At an early age, I vicariously enjoyed his adventures with huge crocodiles, giant pythons, and massive grizzly bears. In retrospect, the dimensions of the animals seemed to increase with the passage of time, so much so that I began to catalog sizes with each retelling. It seems we humans are enthralled with largeness, but the reality is that most dinosaurs were really of quite modest proportions.

Later in life, when I began my studies in biology, whole new battlegrounds and a vast array of new predators became visible to me under the lenses of a microscope. Armadas of swift and slow-moving microorganisms were involved in survival contests. Some seemed to dart across the field at tremendous velocity, but speed is of course only relevant within a given dimension: 30 body lengths in a second for a paramecium is not comparable to the 15 miles per hour of a great white shark. Some of the microscopic warriors that I watched fired salvos of missiles that immobilized their prey, while others gave off toxins and other chemical defenses. Still others used early-warning systems to detect predators and escape. Indeed, these tiny predators had evolved their extraordinary weapons systems over 3 billion years ago, and under selection pressure their prey had evolved a variety of countermeasures to ensure their own survival. It was in this way that predators and prey coevolved.

The evolutionary selection pressures that pit predators against prey in an endless struggle are analogous to an arms race, powered by competition. In this race there is never a clear-cut victor, and advantages that accrue to either of the contestants are simply statistical improvements in efficiency comparable to an improved batting average. A small increase in predator hunting success does not have significant impact on the prey population, and the development of antipredator countermeasures has little impact on the predator population. The agenda for survival of both groups involves a variety of changes in the total weapon system, which includes not only the design of the hardware of actual weaponry but also the tactics and strategies that involve coevolution of both the hunters and the hunted. The evolutionary processes that improve fitness go far beyond adaptations involved in killing and eating and defensive strategies to avoid being eaten. Indeed, predation, in a bizarre way, ensures improved survival of prey species. It is mutually beneficial, because predators usually cull out the weak, the old, and the less fit of the prey species, and in doing so they preserve the dynamic balance of nature.

The environment, or battleground of competition between predator and prey, has an enormous influence on survival, and in the history of life on Earth, the environment has undergone dramatic changes. There have been cataclysmic physical events that led to mass extinctions, invariably followed by extreme environmental changes, which in turn allowed survivors of the extinctions to evolve and adapt to fit into a variety of niches vacated by previously dominant species. However, until the recent arrival of the human species on the scene, seldom has one species caused the extinction of another.

The top predator on the planet, the dominant species, is the social animal called *Homo sapiens*. Humans think, moralize, have consciences, exhibit compassion, and perform acts of kindness; being toolmakers, however, we have also created an armory of extraordinarily lethal weapons. Only humans have instruments of mass destruction, only humans are capable of genocide, cruelty, and the torture of other members of their own species. In the last chapter of this book, I've tried to provide some insight into the paradox of the dark side of human behavior.

It was a daunting task to compress into a book-length story 3,500 million years of the history of the struggle for survival by predators and prey. For this reason, I've limited my focus to a number of familiar organisms, past and present. The intended audience is the general reader interested in both natural and human history. However, the book should also have broad appeal to those in specialized branches of science and technology. The elegance of natural design and the extraordinary technologies of nature—ranging from sensing, data processing, biomechanics, and locomotion—often surpass the most sophisticated and advanced weapon systems devised by humans. Indeed, many of the new strategies, tactics, and technologies involved in human conflict can be found in some of the most ancient, and supposedly simple, life forms.

Acknowledgments

In the process of putting this book together, I have drawn on the accumulated knowledge and insights of many of my teachers, colleagues, and coworkers, who provided me with a veritable treasure-trove of information, ideas, and suggestions. I am particularly indebted to Professor Thomas Kunz, whose collective knowledge about vertebrates, particularly bats and small mammals, added immeasurably to the book. Professor Les Kaufman shared with me his vast knowledge of predator-prey relationships in aquatic environments. Professor Fred Wasserman's assistance on the intricacies of bird behavior helped make the chapter on aerial conflict more exciting. I am also indebted to the enthusiastic support of Professor Eric Widmaier on a variety of topics. These colleagues, along with many others in the biology department of Boston University who let me pick their brains, helped bring this project to fruition.

I would also like to acknowledge a number of authors whose popularized scientific writings provided me with examples of how to write a readable, yet authoritative, book for the enjoyment of a wide public audience. To provide a good read, telling a lucid, compelling story, I leaned on the stylistic approaches of my former colleague at Boston University Medical School, the late Isaac Asimov. I am also indebted to the scholarly yet popular writings of Stephen Jay Gould, whose breadth of knowledge and command of the language I can't attempt to match, but whose skill at clarifying complex subjects involving evolutionary processes helped me immeasurably in weaving the thread of evolution as a unifying theme. May R. Brenbaum provided me with a model of masterful storytelling that was lively, lucid, and at times humorous, and gave me the courage to allow subtle humor in my own manuscript. I would also like to express my appreciation of the writings of William Agosta, from which I gained insight into how to incorporate a lively story that would convey the excitement of life's intricacies to both a general and a scientifically trained audience.

I am particularly indebted to John Michel of W. H. Freeman and Company for granting me the opportunity to take a jumble of 40 years of accumulated scientific information and turn it into a book for the general reader. His patience, encouragement, and enthusiasm for the project were essential in bringing the book to print. I would also like to thank the whole editorial staff, particularly Amy Trask, for their help in the organization of the text and integration of the illustrations and captions.

One of the features that make the text exciting is the exceptional artwork of Trudy Nicholson, a talented, award-winning biological illustrator. Her ability to create dramatic, scientifically accurate drawings helped immensely in the preparation of the book. I had access to over 300 drawings that she had done for the annual calendars of the Grass Instrument Company of Quincy, Massachusetts. Mrs. Grass put me in touch with Nicholson, and I was able to gain permission to use some of these drawings from Astro-Med, Inc., which had taken over the Grass Instrument Company. I would like to express my deepest appreciation to them. Nicholson also developed a number of new illustrations specifically for this book; and in each case she researched the drawings, sometimes even traveling to the American Museum of Natural History in New York to consult with experts on the anatomy of the organism being depicted. Her input went well beyond that of simply illustrating the text, because she frequently had constructive suggestions about the text itself.

Being computer illiterate, I had to rely on a number of people with word processing skills, the most important of whom was Marie Mota of the Boston University biology department. She edited, spell-checked, and made endless corrections, and for this I owe her thanks.

There are several other people and events that helped shape the development of the book. My father, cartographer, explorer, and former naval officer, whetted my interest in nature with tales of his experiences in the jungles of Southeast Asia, and in the navy in World War I. My interest in the military became personal in World War II, when I earned my aerial gunner's wings and navigator's wings and flew as a crewmember in B-17s. With my younger brother, the late First Lieutenant Matthew Levy, a pilot with the tactical air force, I shared many hours of discussion on weapons, strategies, and tactics.

My interest in the history of natural conflict was well established by the time I became a biologist and this interest was augmented by my field experiences as a U.S. Park Ranger Naturalist. Later in my career, I was awarded a Fulbright Scholarship at the University of East Africa in Nairobi, Kenya, which gave me the opportunity to see unspoiled wildlife in the forests and savannahs of Tanzania, Uganda, Kenya, and Ethiopia.

Finally, I want to acknowledge the efforts of my wife, June, who was delighted to learn that her pet name, "Junebug," was also the name of a fearsome predator of plant pests. She enthusiastically supported my work, encouraged me in my toils, and tolerated my long hours of unavailability. Her enthusiasm for life, her humor, and her intelligence have made my writing ever so much better.

Chapter 1

In the Beginning

The Origins of Earth

About 4½ billion years ago, a ball of cosmic debris coalesced to form our planet some 93 million miles from its parent star, the Sun. This distance is fortuitous because it means that Earth is in the liquid water zone, far enough from the Sun so that water does not boil and close enough so that water does not totally freeze.

Early Earth was a molten ball, heated by the energy released in the decay of radioactive material and meteor impacts. As it cooled, Earth divided into layers: The inner and outer molten cores, made up of heavy metals such as iron and nickel; the lower mantle (1,700 miles thick); the upper mantle (400 miles thick); and the outermost layer, the rocky crust (6 to 25 miles thick). Gases expelled in the process of Earth's formation created a primitive atmosphere very different from that found today, an atmosphere nearly devoid of oxygen.

Being composed of the lightest materials, Earth's crust literally floats on the more dense lower layers of the mantle, which circulate, driven by heat released from the core. The surfaces that make up the continents are driven by this motion, moving as vast plates in a gigantic recycling engine: new hot materials push up from below, while at the margins, old crusts cool. When the moving plates collide, the cooler material is pushed back into the core; in areas called subduction zones. So, crust is constantly being created and destroyed. The movement of the plates is slow. For example, India and Australia, pushed apart by hot plumes rising from the mantle, split off from the Antarctic plate; it took 150 million years before India smashed into the Asian plate, causing a massive buckling up that formed the Himalaya Mountains. (In fact, India is still moving northward into Asia and the Himalayas are still growing.) Over time, the constant movement of plates formed supercontinents, which rifted apart and came together and rifted apart again to form the present continents. During the past 650 million years, continental drift spawned monumental changes in Earth's atmospheric temperatures. There were alternating hot and cold periods, during which huge ice caps formed and melted, and ocean levels fell and rose. The rifts in the crust also led to immense volcanic eruptions which had dramatic effects on Earth's environment. In all probability, these eruptions played a major role in some of the great extinctions of life on Earth.

Vulcanism and Earthquakes

As heat, escaping from the mantle, rises in thin "roots," it reaches the bottom of Earth's crust and spreads out, forming a dome of hot molten rock, or magma, lifting the crust, and fracturing it. Gases are released from the magma, sometimes with such explosive force that the crust is shattered into tiny dust particles, and red-hot molten lava flows out over the surface. In the course of human history, we have recorded a number of large, catastrophic volcanic eruptions and severe earthquakes caused by plate movements, but these pale in comparison to the incredibly violent events of the past.

About 250 million years ago, a time period that coincides with the mass extinction at the start of the Paleozoic period, a series of massive eruptions occurred in what is now Siberia. A gigantic flood of lava, over 700 miles of it, poured out, and vast quantities of dust, droplets of sulfur dioxide, and other gases were blasted into the atmosphere. These substances spread out over Earth's entire surface and caused both short-term and long-term effects on its environment. The largest volcanic eruption of recent history, which occurred in 1815 on the island of Tamboura, in the East Indies, pumped enough ash and gas into the air to diminish the amount of sunlight hitting Earth's surface and lower temperatures. This effect was substantial enough that in Europe it was called the year without a summer.

During the eruptions in Siberia and the later massive volcanic eruptions in India, mind-boggling quantities of material were injected into the atmosphere: trillions of tons of carbon dioxide, trillions of tons of sulfur, billions of tons of flourine and chlorine, and untold amounts of volcanic ash. The first effect was to shield the earth from sunlight and cause the death of many photosynthetic organisms, organisms that normally sop up carbon dioxide (CO_2). Earth's temperatures dropped rapidly, by as much as 10°F. Sulfur dioxide reacting with water vapor produced clouds of acid, which subsequently fell to Earth's surface, acidifying the seas and killing oceanic photosynthesizers. Still loaded with CO_2 and other gases, the atmosphere later reflected heat rising from the surface back toward Earth—a process similar to what happens in a glass greenhouse. As a consequence, average global temperatures subsequently rose as much as 10°F.

Along with the rifts and fractures of the crusts, there were earthquakes of staggering proportions, some so severe that they would reach 13 or 14 on the

Richter scale, 1 million to 10 million times greater in magnitude than anything in recorded history. The cataclysmic eruptions and devastating earthquakes that occurred in or near the oceans caused tidal waves, or tsunamis, that sent walls of water over 200 feet high smashing inland over low-lying coastal environments. Certainly, such physical changes on Earth can account for some of the mass extinctions that occurred in the past 600 million years. But there is another form of severe physical disruption that could cause even greater effects: the impact of some reasonably large extraterrestrial body (an asteroid or comet) on Earth.

The Extraterrestrial Impacts

From the very beginning, Earth has been bombarded by extraterrestrial objects: comets, asteroids, and meteorites. Just a quick glance at our moon through a small telescope reveals a massively pockmarked surface, covered with large and small impact craters. On Earth itself, we find scars left by large extraterrestrial objects that passed through our atmosphere without burning up and smacked into Earth with devastating force. A large meteorite hit Arizona just 3,500 years ago and created a bowl-shaped crater almost a mile across. Even as recently as 1908, a comet or meteorite a few hundred feet in diameter exploded over Siberia, leveling every tree in over a thousand square miles. The object, traveling at about 35,000 miles per hour on impact, generated local temperatures of about 30,000°F. What would the effect be of even larger extraterrestrial impacts on Earth's environment?

Recently, evidence of another large impact, also in what is now Siberia, has been dated at 35.5 million years ago. The extraterrestrial object left a crater, known as the Popogai structure, 100 kilometers in diameter. The time frame of this impact, at the boundary of the Eocene and Oligocene periods, coincides with the most severe extinction since the demise of the dinosaurs. We can correlate this impact with the cooling of the oceans and the appearance of ice sheets over Antarctica. At about the same time, another impact left an 85-kilometer-wide crater in the area of the Chesapeake Bay on the east coast of North America. Thus, two successive cataclysmic impacts are implicated in the mass extinctions of the Eocene-Oligocene boundary.

Recently, the site of the impact of a mile-wide asteroid in western Argentina has been uncovered by a team of scientists led by Peter Schultz of Brown University and Marcello Zarate of Argentina. The crater is thought to be off the coast and has yet to be located. But glass fragments created by the heat of the impact are rich in iridium, a material rare on Earth but common in asteroids. Sediment cores of the nearby seafloor show that there was a sudden drop in temperature about 3.3 million years ago that lasted about 50,000 years, possibly triggering the beginning of an ice age. It has been suggested that this impact was responsible for the demise of the giant marsupial ground sloths; the monstrous armored armadillos; the 8- to 10-foot-tall, flightless, carniverous terror birds; and other huge mammals that became extinct after the species-killing impact.

Most of us have read somewhere in the popular press about a similar event that occurred about 65 million years ago and wiped out the dinosaurs. There is now overwhelming evidence that an asteroid, 6 to 9 miles across, weighing a trillion tons, and traveling at a speed of about 150,000 miles per hour, smashed into Earth on the Yucatan Peninsula. The impact created a crater about 30 miles deep and 90 miles in diameter. The explosion injected such a mass of dust and water vapor into the upper atmosphere that no sunlight penetrated it for a number of years. The heat of the impact and the hot, falling debris started massive forest fires. In the area of the impact the intense heat caused atmospheric water to react with nitrogen and oxygen to make nitric acid, which then precipitated out. The precipitation of the acid changed the acidity of the oceans and killed most species of aquatic microorganisms. Later, because of the lack of sunlight, Earth's land surface temperatures fell to below freezing. Later still, the accumulated atmospheric carbon dioxide increased fivefold, producing a greenhouse effect, raising global temperatures more than 10°F, disrupting food chains, producing cataclysmic climate changes, and annihilating more than 50 percent of the species of plants and animals then dwelling on Earth, including the dinosaurs.

Relating the mass extinctions revealed by geological studies to catastrophic impacts or episodes of monstrous vulcanism is difficult, and controversy rages between the proponents of different theories. We have compelling evidence of six great extinctions and several smaller ones that occurred from the Cambrian period up to the modern era. Thanks to great technological advances in radioactive dating of rocks and sediments, the time of these cataclysmic events is fairly well pinpointed, but the exact cause or causes of the massive annihilations are still a matter of very heated debate. The details of this debate are clearly presented in a very readable fashion by Lowell Dingus and Timothy Rowe in their elegant book, *The Mistaken Extinction, Dinosaur Evolution and the Origin of Birds* (New York: Freeman, 1998). However, the evidence for the cause of the mass extinction at the end of the Cretaceous period overwhelmingly points to an impact by a very large asteroid 65 million years ago.

Although each extinction killed off varying numbers of plant and animal species, it also left survivors to fill the ever-changing niches created. These survivors multiplied, evolved, diversified, and took over.

Ten Minutes to Extinction   A trillion-ton asteroid heads for the Yucatan about 65 million years ago. Traveling at 150,000 miles per hour, it produced such devastating long-term environmental changes that a large number of terrestrial and aquatic plants and animals went extinct, most notably the dinosaurs. These changes signaled the end of the Cretaceous period and the beginning of the Tertiary period.

The Nature and Origins of Life

Ever since the human species became aware of itself, it has been concerned with its nature and origins. Indeed, as one studies the fossil records of our early ancestors and the myths of Stone Age peoples, evidence of our ancestors' interest in their own nature can be seen as a measure of their progress toward humanness. Their early speculations led them to realize that they were part of a large group of Earth-dwelling objects that were alive; so, their first attempts to understand the world probably involved the separation of living objects from nonliving objects. Once they recognized differences between animate and inanimate objects, humans developed an interest in the nature of life itself and how it began.

Only in this century, and particularly in the last fifty years, however, have we acquired enough understanding of chemistry, physics, and astronomy to replace mystical speculation with more precisely defined concepts of chemical reactions, molecular structure, molecular interactions, and a variety of biophysical events.

With our expanded knowledge about the chemistry of life and our increased understanding of information coding at the molecular level, we have been able to create scientifically sound theories about how life evolved from nonlife, what produced diversity, and how more-advanced forms of living things evolved. Not only are we able to create theories; we are also able to devise laboratory experiments to test some of these theories. A number of these experiments have demonstrated successfully that the constituents of the primeval nonliving Earth could, under proper conditions, be converted into more complex combinations of atoms that are similar or identical to molecules found in living organisms.

What Is a Living Organism?

Living is a very difficult adjective to define. Any attempt to define life must be somewhat subjective and arbitrary. But we humans are very arbitrary beings; we love to make long lists of what is acceptable as a condition for life (and almost anything else). Still while we like sharp, clear-cut separations of living from nonliving, we are rational enough to realize that over the vast reaches of time, there was undoubtedly a gradual transition from what is clearly nonliving to something that *may be* living to something that is clearly alive. Simply to avoid argument, Nobel laureate Melvin Calvin came up with two qualities that everyone could agree on as basic characteristics of a living system. These were (1) "the ability to transfer energy and transform energy in a directed way" and (2) "the ability to remember how to do this, once having learned it, and to transfer, or communicate, that information to another system like itself which it can reconstruct."

What do these two qualities mean? The ability to "transfer and transform" energy signifies that living things are able to convert chemical and physical energy into energy that is useful to themselves, photosynthesizing sunlight, for example, or using food to fuel their bodies. The sum total of chemical reactions that either build new molecules or break down existing molecules to transform their chemical energies is referred to as metabolism. All living things are capable of metabolism. The communication of information implies that living things are responsive to their environment and can reproduce their own kind. In other words, living organisms have the capacity to use their information-containing molecules, not only to produce similar organisms but also to program the functioning of those organisms within their environment.

Salvador Luria, also the winner of a Nobel Prize, points out that living is distinct from all other natural events in that it has a program, whereas physical phenomena are essentially random events tending toward increasing disorder. Life not only manifests individuality but also in a complexity maintains order not otherwise seen in the physical world.

The Program for Life

The material that contains the blueprint for life, the programming substance that has persisted in a variety of forms for over 3½ billion years, is the gene. The gene is a molecule whose very construction ensures stability and, thereby, continuity in the general features of any species or plant or animal group. At the same time, the structure of the gene can be altered to produce a vast variety of specific changes, changes that permit evolution to take place. So, another characteristic of life is its ability to evolve.

Evolution is not seen in the physical world. Crystals reproduce, but this repetitious process always produces identical crystals. In contrast, because living things are capable of change, they have produced diversity unheard of in the nonliving world. But somehow life must have begun as a chemical, that is, a *physical* phenomenon; life must somehow have arisen from nonliving components. While we have no direct evidence about how and where the first gene-like, self-replicating molecules appeared, we can speculate about how they came to be.

How It All (Probably) Began

Life is an historical process of which we have a very incomplete record, and we certainly have no record of life's origins. Compare, for a moment, our study of life's origins with our study of evolution of the later-living forms. We know that the history of life represents a selective sequence of events: some living organisms were successes, and their descendants are the ancestors of today's living forms. But, as a 3½-billion-year process of trial and error selected winners and losers, more forms became extinct. Life has produced many more failures than successes, and we know about many of these failures because they left fossil remains. Unfortunately, however, the earliest failures in the transition from simple molecules to life left no trace. Nevertheless, we can look back in time and speculate about how it all began.

Primitive Earth, the Sun, and all the other stars in the universe are chemically similar, containing a number of elements common to living organisms: hydrogen, carbon, oxygen, nitrogen, and phosphorous. These elements may become linked by chemical bonds to form molecules. Molecules may be simple combinations of two atoms of any element, or they may involve a million atoms forming huge complex molecules, or macromolecules, such as the giant informational molecules that were necessary for life to begin.

The step from simple molecules to macromolecules required millions of years and initially produced some very simple molecular constructions: carbon or nitrogen atoms bonded to hydrogen atoms, carbons bonded to nitrogens and oxygen, oxygens bonded to hydrogens (water), and phosphorous bonded to oxygens. The early atmosphere of Earth had in it just such bonds in the form of water, carbon dioxide, methane, molecular hydrogen, and ammonia. These primitive molecules were to be the initial building blocks from which progressively more complex molecules would form, but something else was also needed. In order for new chemical bonds and new combinations to be created, there would have to be a source of energy.

Then, as now, the major source of energy was the Sun. Primitive Earth had no protective atmosphere to absorb much of the radiant energy coming from the Sun. Some of that energy was in the form of X rays and other cosmic rays and some was in the form of visible light that could be converted to heat, but most was in the form of ultraviolet (UV) light. Thus, in a single year during this early era, 35×10^{20} (3,500,000,000,000,000,000,000) calories of ultraviolet energy reached Earth's surface. Other massive energy inputs were provided by electrical storms, heat, and radioactive decay of materials in the Earth's crust. All of these energy inputs were necessary to create the molecules that were to be the basic building blocks of life.

For energy to have an effect on a molecule, it first must be absorbed. Then the molecule becomes excited and enters a highly unstable reactive state. In this condition, it is more likely to combine with other molecules. The transformations of the first building blocks to more complex molecules were powered by an abundance of energy.

The Cryptogram

The blueprint of life's secrets was written on a primitive molecular papyrus in the form of an ordered array of four molecules, adenine, guanine, cytosine, and thymine, which symbolically constituted the alphabet of the language of genes. Trios of the four letters were in turn coded for the 20 amino acids that make up the alphabet for proteins, the working molecules of life. With these codes, the early genes could not only reproduce themselves but also dictate instructions to make and maintain living things. There is good evidence that the first code, or cryptogram, ribonucleic acid (RNA) was used by the early living organisms that formed in the scalding environment of volcanic hot springs on the primitive oceans' bottoms. Loaded with dissolved metals and other energy-rich atoms, these hot springs were kitchens in which all the stuff of life was cooked up. However, a brew of life's constituents is still not alive. It had to be packaged within a living container, a membrane that could selectively let materials in and out. It is believed that the first membrane-bound living organisms appeared a bit over 3.8 billion years ago as simple, microscopic bacteria-like organisms. These were the ancestors that paved the way for all the more complex forms yet to come.

The early, heat-loving bacteria gave rise to several other branches in the tree of life. One main branch was the bacteria that supply energy for all plants and animals. Later, the primitive archaebacteria branched off, as did the ancestors of modern bacteria and blue-green algae. Some of these early bacteria modified their ability to get energy into trapping the Sun's energy, using water, and fixing carbon dioxide. In so doing, they released oxygen. This process, photosynthesis, was critical in the evolution of more-advanced true cells that contained their own advanced genetic material, a damage-resistant double spiral of deoxyribose nucleic acid, DNA. The addition of oxygen to the environment changed the world dramatically; it altered metabolism and accelerated other life processes. New species of oxygen-using bacteria, the aerobes, appeared. While oxygen was a poison to many of the primitive bacteria, it was essential to all higher organisms, single-celled protists and multicelled fungi, plants, and animals.

Thanks to the tireless and inventive work of renowned microbiologist Lynn Margulis, we have insight as to how these early bacteria played a role in the evolution of higher life-forms. According to her theory, photosynthetic bacteria were incorporated into other cells and developed a mutualism, or symbiotic relationship. Photosynthesizers became miniature organs, chloroplasts, that provided the host cell with carbohydrates. Chloroplasts retained some of their own DNA and still had the capacity to reproduce. As more and more photosynthetic organisms populated Earth's waters and land masses, oxygen levels rose, reaching a peak level in the Carboniferous period. (Today, at sea level, our atmosphere contains 21 percent oxygen, and it is calculated that the current level of photosynthesis replaces all atmospheric oxygen every 200 years.)

Margulis's theory also helps us understand the origin of oxygen-breathing organisms, those organisms which get their fuels by ingesting other organisms. These fuels have tremendous amounts of energy stored in their chemical bonds, but organisms cannot use this energy directly. First, they must break the fuels down and then convert that energy into usable, immediately available energy. To accomplish this, they must use processes whose last step is always the addition of oxygen. These processes are provided by mitochondria, resident ancestors of aerobic bacteria that evolved as symbionts, just like the chloroplasts. These powerhouses of the cell have their own DNA and can reproduce themselves. Some cell types contain up to a thousand mitochondria. Whereas processes in a cell lacking oxygen produce only 2 usable, high-energy molecules per molecule of sugar, aerobic processes produce 36 such high-energy molecules! Armed with a new, highly efficient metabolic machinery, the advanced single-celled organisms evolved to become predators and scavengers that could exploit the energy-rich nutrients found in other organisms. About 2 billion years ago, the protists in the ancient seas experienced an explosion of new and diverse forms. These forms were both hunters and hunted; the arms race had begun.

The Arms Race Begins

As more-advanced bacteria evolved, some became motile by gliding along surfaces. Others became swimmers, using long, oarlike appendages, called flagella. They also developed a primitive armor, a nonliving cell wall that was intended, primarily, to protect them from the harshness of their environment. Although there weren't many predators, there was still competition for energy resources and space. However, there was no competition for mates, since the bacteria were sexless and reproduced by simple fission after duplicating their hereditary material. Eventually, some bacteria developed a primitive form of chemical warfare. They were able to synthesize and release toxins which killed other bacteria adjacent to them, ensuring that their energy supplies would not be usurped. Three billion years later, humans were to discover the antibacterial toxins (over 500 distinct antibiotics are produced by the streptomyces alone) for the treatment of human-disease-causing organisms.

While most bacteria derived their energy from decomposing debris left by others, it was only a matter of time before some mutant occurred that couldn't wait for something to die but would instead seek to get its nutrients from other living bacteria. We have no evidence of when this happened, but looking at the rare bacterial predators of today, we can see some small bacteria, the *Bdellovibro,* preying on larger bacteria to exploit their constituents as a food source.

Predators have to be quick, and the flagellated vibrios were that. Predators also have to be able to latch onto their prey (*Bdello* means "leech"). The vibriod penetrates the prey's armor and feeds on its nutrients; it then reproduces itself. There are only a few such examples of predatory bacteria. They have been given the colorful name vampirococci, the vampire bacteria.

In all probability, the early bacteria were also victims of a wide range of viruses called bacteriophages. The name *virus* originally referred to any form of poisonous envenomation. In terms of living things, viruses fall into a gray area; they are incapable of metabolizing nutrients needed for their reproduction, and in order to reproduce, they must invade another cell and take over its genetic and metabolic machinery. Sometimes the invasion of a host is destructive, and the host cell bursts, spilling out hordes of new virions, or infective viruses. Since they lack a locomotion mechanism, they are incapable of searching for their prey; instead, they float freely and rely on random chance encounters with a host. Clearly, the odds of finding a proper host are enhanced by increasing the number of viruses, and so viruses typically reproduce enormous numbers of themselves.

Outside of the host, the submicroscopic (about a millionth of a millimeter) virions, though metabolically inert, do have structure; a protein outer envelope, or head, contains within it a genetic blueprint capable of taking over the host cell's genetic machinery. The most-complicated viral structures are found in those viruses which attack bacteria. They look like miniature lunar-landing modules with leglike fibers that can latch onto a specific, susceptible bacterial host. After attachment, the phage uses a penetrator, a hollow, hypodermic-needle-like mechanism, to inject its own genetic program into its victim. Once inside the host, the viral DNA hijacks the host's metabolic machinery. Acting as a sort of hereditary fifth column, the virus takes over, and its own genetic material and proteins are reproduced. These subunits are then assembled and packaged as many new viruses, which are released when the host cell lyses or bursts.

The arms race began to heat up a little over 2 billion years ago when the first true cells appeared. They had the advantage of having internal membrane-bound structures that enhanced their capacity to rapidly metabolize nutrients and synthesize new components. They also had their genetic program encapsulated within a porous double membrane and had a much more sophisticated hereditary apparatus than their ancestral algae and bacteria. Over the next 2 billion years, they evolved into a myriad of organisms. They were called protists and were much larger and faster than the bacteria. Initially, the predatory protozoa browsed on bacteria. Some, like the amoeba, could recognize, flow around, attach, and engulf prey. Others were propelled by the new and vastly improved flagella, and some developed short, flagella-like structures, or cilia, which beat under the control of a synchronizing mechanism that allowed them

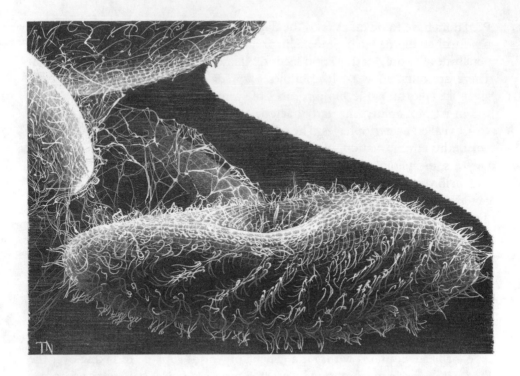

An Early Propulsion System   Single-celled paramecia move through their watery surroundings by coordinating the movement of the oar-like cilia that cover their surface. They can change direction and speed quickly to avoid predators or in response to a variety of chemicals in their environment.

to move rapidly both forward and backward. While some cilia were used for locomotion, other ciliated structures acted as a primitive gullet, creating a current which swept the prey into a feeding chamber for digestion.

How Fungi Make War

About 600 million years ago, one group of ancestral protists spawned a group of organisms superbly adapted for absorbing nutrients, the fungi. Most of them made their living by degrading dead remnants of other organisms. During the subsequent diversification of fungi, some became parasites on other organisms. One, *Amoebophilus*, specialized in attacking single-celled amoeba. Others specialized in preying on and parasitizing many diverse organisms, including other fungi. It is not surprising that with the evolution of humans, a variety of fungi became human parasites.

The fungi consist of a body, or mycellium, that can organize its cells into rapidly dividing filaments, or hyphae, which in parasitic forms, are modified to penetrate the cells of the host. The fungi thrive in diverse environments; some have adapted to the role of active hunters whose hyphae secrete sticky, gluelike chemicals that immobilize passing creatures. Once the prey is trapped, the hyphae invade the prey, spreading rapidly, absorbing its components, and eventually killing it. Another remarkable fungal weapon system evolved in soil fungi, whose habitat is rich in tiny roundworms (nematodes). The weapon system of soil fungi is highly adapted for these particular prey. Soil fungi form three cell rings that match the diameter of their prey. When a nematode crawls through a ring, it causes the ring to swell and, literally, lasso the worm, which is then invaded and digested.

The Biological Big Bang

Life processes are a continuum. Ancestral genetic blueprints are modified and then slowly tested in the sieve of natural selection, all within an ever-changing environment. However, every now and then, the fossil record shows a rapid transition from previous body plans to a new and rich diversity of life-forms, which seem to have suddenly appeared on the scene. One of these transitions, the so-called Cambrian explosion, began about 530 million years ago. In a relatively short time, a few million years, a multitude of new multicellular organisms began to fill the seas: clams; snails; jointed, legged animals (arthropods); trilobites; echinoderms; and even primitive chordates, the ancestors of all vertebrates, including humans. Many of the creatures left a record of their existence in the marine sediments, and many were truly bizarre. Most were wiped out in a huge mass extinction about 40 million years later. This abrupt transition saw alterations in body plans of disparate architectures. So there had to have been a major genetic development—a genetic regulatory program that specified the activities of genes in different cells, which expressed themselves in the subsequent development of the organism.

Development is carefully regulated by a series of time-dependent control switches. The process can be compared to a row of dominoes falling over. The first fall initiates the second fall, the second initiates the third, and so on. In genetic terms, the genes express themselves by making proteins, and these proteins, in turn, switch on other genes, producing a cascade of events early in the development of an organism. Such genes set up a basic body plan, dictating instructions for a head-to-tail axis and a dorso-ventral (back-belly) axis. Later on in the cascade, various major body parts, such as head, thorax, and abdomen, are blocked out, and later yet, more-refined structural details, such as mouth parts, eyes, antennae, appendages, and segments, are set up. As new genes express themselves, older genetic programs are switched off. We now know that many classes of regulatory genes share a common informational sequence. This sequence, known as homeobox (HOX) genes, arose just before the origin of complex multicellular animals. Obviously, such a highly sophisticated regulatory system was not needed by our single-celled ancestors which reigned supreme for some 3 billion years. Our simplest multicellular ancestors, the sponges, had only 1 HOX gene. The arthropods (crabs, insects, spiders, etc.) have 8, and humans have four clusters of the 38 regulatory genes. But, the HOX genes are only part of the story. The great diversity of structure seen in each group of animals is generated by other genes that are activated only after the HOX genes provide the basic format. A detailed discussion of the discovery, as well as the action of HOX genes is well beyond the scope of this book, but it is comparable to the discovery of the wheel, which led to the diversity of vehicles seen in the very brief history of humans.

The Bizarre Creatures of the Cambrian Big Bang

The cause of the Cambrian big bang is unknown. Most evolutionary biologists think it was related to some pervasive change in the environment of that time, possibly an increase in atmospheric oxygen. What is clear is that it led to an arms race in which predators and prey developed the elaborate weaponry and sophisticated strategies and tactics that are carried on by their descendants today. The suddenness of the appearance of these animals with eyes, tentacles, claws, jaws, spines, and armor is unprecedented in the history of life. It all happened in a span of about 10 million years, a mere blink in the evolutionary course of time.

Creatures of the Biological Big Bang (facing page)   Fossils found on the 530 million-year-old Burgess Shale reveal that in the relatively short span of a few million years, an enormous number of new and bizarre life forms appeared—an evolutionary big bang of sorts. Pictured here is a view of what some creatures from the Cambrian Seas looked like. The large bug-eyed animal at the top is a three-foot long predator called *Anomalcaris*, and just below it (on the left) is the five-eyed *Opabina*, equipped with a grasping proboscis. Just below and to the right of *Opabina* is *Picaia*, possibly the oldest known ancestor of the vertebrates. Walking on the ocean bottom at the lower left is *Hallucigenia*, with seven pairs of spines and tentacles. At the bottom right is the harp-shaped *Marella*, equipped with especially long antennae.

The remnants of the big bang, many long vanished from the planet, are engraved in slices of shale excavated from geological formations in the Canadian Rockies (the famous Burgess Shale) and China. There were wormlike ambush hunters that burrowed into the seabed floor capable of extruding a thorny proboscis to snag their prey. There were arthropods that had sensory antennae and eyes on stalks; some had spine-tipped appendages, while still others had clawed legs. There were flexible plates of armor and defensive spines. Many were predators that preyed on the abundance of simpler multicellular organisms that preceded them. Some, such as *Opabinia,* had five eyes; segments covered with armor; and a long, flexible proboscis with clasping jaws. They were like creatures out of science fiction. One, so bizarre that it seemed to be the product of an hallucination, was promptly classified *Hallucigenia.*

Mass Extinctions and Radiations

Examining the 500-million-year history of complex multicellular animals reveals six massive extinctions. While the cause of most of these extinctions is not clear, we do know that each was followed by a radiation of surviving lifeforms. Such radiations give rise to new variations, which eventually become diverse new species.

One of the most dramatic of these radiations took place between 440 and 510 million years ago on the warm, sunny tropical seas. Called the Ordovician radiation, it produced organisms that would dominate all the oceanic ecological niches for 250 million years. But the Ordovician radiation is unusual in that it was not preceded by a major extinction. The more usual course of radiations is that they follow a cataclysmic event, and the survivors of the dramatic change then exploit the opportunities made open to them by the demise of previously dominant species. This was how, after the great extinction of flying and marine reptiles and of the dinosaurs, the birds and mammals rose to ascendancy. For example, accumulated recent evidence for the demise of the dinosaurs, many marine species, and all of the ammonite mollusks, about 65 million years ago, points to a massive meteor impacting the earth in an area near the Yucatan Peninsula. Whether other extinctions were caused by extraterrestrial bodies hitting the earth, by vulcanism, or by a combination of both remains unknown.

Strategies for Survival

The survival game between predators and prey involves an often-repeated contest, in which, if the predator wins, its payoff is the acquisition of nutrients and the termination of the prey. It should be added that the winner, in this case, only wins if the payoff in the calories taken in exceeds the calories expended in the hunt. On the other side of the game, the prey wins if its defenses and escape strategies deter the predator and it lives to fight another day. In such games, there are only winners and losers. But there are other strategies in survival games; one in which two members of the same or different species enter into a joint effort to the mutual benefit of each other is called a win-win strategy. Examples of such symbiotic relationships go back to the earliest life-forms. As we saw earlier, mutualism played a major role in the evolution of both plants and animals, by providing them with chloroplasts for photosynthesis, as well as mitochondria for aerobic metabolism.

Other symbiotic relationships evolved and are still active today. For example, wood-eating termites have the apparatus to cut up wood and swallow it, but they cannot digest the energy-rich cellulose without the aid of symbiotic microorganisms living in their bowels. These organisms have enzymes that can split the bonds between the sugar molecules that make up cellulose. The sugars thus produced provide the termite with its energy supply, and the termite provides the symbionts with housing and food. Similarly, in the human large intestine, there are many symbiotic bacteria, paying rent in the form of some vitamins essential for human survival. Such win-win strategies would seem to be ideal for survival, but both participants in the game must accept the rules of play.

If one of the players in such a game chooses to defect and tries to get without giving, a mutualistic relationship becomes a parasitic one. In an evolutionary sense, parasites are former mutualists who cheated and, in so doing, benefited at a cost to their host. This strategy has proven to be a very successful one: in today's world, there are twice as many parasitic species as there are free-living species.

Survival Requirements of Hunters and Hunted

For mobile predators and prey, success in preserving self is often measured in fractions of a second. In this game of survival, response time depends on several systems. The first is an early warning system that can detect the antagonist. This type system is composed of specialized sensory cells that can recognize environmental cues that are either physical or chemical, for example, touch, pressure, light, vibration, odor, electric field, and so on. Over the course of time, natural selection resulted in ever more efficient sensors, some of which are truly extraordinary and can't be duplicated even by the advanced technol-

Survival Strategies   The sea butterfly, *Clione*, is an unarmored sea snail that normally floats head up. When threatened, it stops moving and withdraws, but when its head makes contact with one of its favorite prey, it goes on the attack. It immediately protrudes its tentacles, orients itself toward the prey, and activates its swallowing response. This complex hunting behavior involves coordinating several different motor responses, implying a sophisticated mechanism for such a simple animal.

ogy of modern weapon system designers. All sensors have thresholds, that is, a minimal signal detection limit, but this is superimposed over a background of noise, which is also picked up by the sensors. The fitness of a sensor can be improved by increasing the signal and minimizing the noise; over time, sensi-

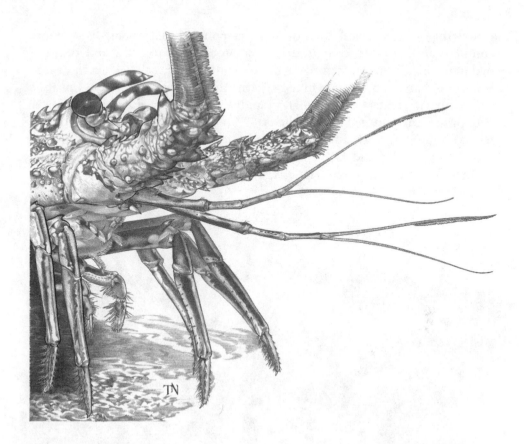

Chemical Cues Underwater   The spiny lobster is both a scavenger and a predator. Although it is armored, it lacks the defensive weaponry of its cousin the northern lobster—large pincer claws—and therefore spends most of the day hiding in the protective shelter of reefs. At night, it forages for food on open sandy bottoms using its enormously long antennae, which flick every second to expose hundreds of odor-detecting sensors to chemical cues in the water. These sensors can not only identify the target but also provide the lobster with a keen sense of the direction of its prey.

tivity has been enhanced. Sensors also have a dead time, immediately following triggering, during which stimuli are not detected. So, natural selection favored sensors that could minimize dead time too. But in developing a defense system, being able to rapidly detect opponents with ever-greater fidelity is only the beginning. The information provided by the sensors must be processed and produce a response. Thus, a sensor must communicate with the body that owns it in order to produce an appropriate survival response.

Early sensory response systems were organized into relatively simple information-processing circuits. These circuits were essentially command centers producing a stereotypical fight-or-flight response. Later systems were more complex and involved complicated data processing, analysis, and decision making, which were performed by the nervous system. In effect, each organism carried an on-board computer. Clearly, the organisms that could process data with the greatest efficiency and rapidity had a survival advantage. Faster data processing required three abilities: (1) to pass information along a circuit more rapidly, (2) to transmit that information between the units making up the circuit rapidly, and (3) to conduct and transmit that information to an "effector." In predator-prey relations, the usual effectors are muscles used in rapid propulsion.

Escape Behavior—Rapid Response Systems Getting away from an approaching predator doesn't allow a lot of time to calculate risks and make decisions. So the best evolutionary strategy is to keep it simple. Thus many fish have specialized escape "programs" hardwired in their brains so that there is minimal data processing.

Speed and Movement

Even before the Cambrian explosion of diverse, multicellular advanced organisms, a new and highly efficient form of motor equipment appeared. It was muscle attached to skeletal structures that were jointed and could act as levers. The late Nobel laureate Albert Szent-Gyorgi conveyed his excitement about muscle: "Muscular contraction is one of the most wonderful phenomena of the biological kingdom. That a soft jelly should suddenly become hard, change its shape and lift a thousand times its own weight, and that it should be able to do so several hundred times a second, is a little short of miraculous. Undoubtedly, muscle is one of the most remarkable items in nature's curiosity shop."

The Cost of a Quick Getaway   Both predators and prey can use up a great deal of energy in movement. Aerial hunters use very little energy when gliding or soaring, but active flight is even more expensive than swimming, and runners expend the most energy. For all types of locomotion, large animals generally use up less energy than small ones. Some animals may perform brief high-energy bursts, such as the rabbit bolting from the predator or the trout propelling its entire body out of the water to catch an insect.

Muscle Power   Some animals are capable of very rapid acceleration for short distances, but these sprints cost a lot in terms of energy and they require specialized muscle fibers that contract rapidly. Because these so-called fast-twitch fibers tire out quickly, animals that move for long distances use slow-twitch fibers, which are long lasting. Of course, most animals' muscles have both fast-twitch and slow-twitch fibers, and it is the relative number of those fibers that determines their style of hunting and escape.

We have learned a great deal about muscle since Szent-Gyorgi's time. First came our understanding of the mechanism of contraction and how it is coupled to electric excitation. The sudden release of calcium from within its storage depots in muscle cells allows the contractile proteins to interact, as energy-rich molecules provide the power for contraction. Movement is expensive in terms of energy, and muscle can store only a minuscule supply of that energy. For contraction to continue, muscle must provide new energy continuously via two metabolic processes, one using sugar as its fuel without benefit of oxygen (anaerobic) and the other a slower process requiring oxygen (aerobic). Both make the fuel that muscle can use directly. But contraction is fuel inefficient; less than 40 percent of the energy available is used to contract and relax muscle. The rest is turned into heat.

There are three major kinds of muscle fibers in all skeletal muscles. First there are the fast-contracting anaerobic fibers, known as white muscle. These fibers generate force to accelerate and reach maximum speed, but they tire quickly. Second there are fast, aerobic fibers (red muscle) which have oxygen-carrying molecules called myglobin. They contract rapidly and don't fatigue as quickly. Finally, the third type of muscle, called slow-twitch fibers, contract and relax slowly; they use oxygen, contain myglobin, and are fatigue resistant. Animals that sprint and accelerate rapidly have relatively more fast-contracting fibers, while animals that can maintain speed over considerable distances have a preponderance of less fatigable, slow-contracting fibers.

If we look at a world-class human sprinter who can go from a stationary position and move 100 meters in about 9 seconds (about 27 miles per hour), we find that in the first 2 seconds, he has used up all his muscles' available energy-rich molecules. In the next 2 seconds, there is a molecular quick fix that reenergizes the energy-storing molecules a bit. To cover the last 5 seconds of the dash, both anaerobic and aerobic processes are working at maximum capacity. By the end of the sprint, he has depleted the muscles' energy stores and is gasping for breath. It will take many minutes to refurbish the muscles' fuel supply.

The fast-contracting fibers also release a by-product, lactic acid, which must be broken down. The fastest land animal, the cheetah, has only enough energy stores to continue pursuit for less than a minute. Such speedy predators are usually lightweight, so they pursue small prey. They have to kill, eat fast, rest, and then hunt again.

With contracting muscles, another problem is how to get rid of the heat that has built up. Cooling takes time, so a predator has to make a cost-benefit decision. Obviously, it is not worth the energy expenditure to pursue the prey in a situation where the chase has a low probability of success. Usually, the hunter estimates the distance he must pursue and how fast the prey will move in its effort to escape. Big, predatory cats, lions and tigers, are heavily muscled and big-boned and cannot maintain a long chase. They must use their strength in stalking and getting as close as possible before attacking. They are, literally, ambush hunters that explode from their place of concealment in a brief, high-speed charge. The prey understands the game, and once he senses the predator's presence, he adopts an alert posture and, to make his awareness even more obvious, will even move toward the predator. In essence, he communicates with the hunter, saying, "I see you, and if you move toward me, I have a big enough lead and am fast enough that you will not be able to catch me." At that point, the predator usually gives up the hunt and seeks a new target.

Elastic Components of Movement

Each muscle fiber, each bundle of fibers, and the whole muscle itself are surrounded by an elastic covering. All of these elastic coverings come together at each end of the muscle and form strong, stretchy tendons that attach the muscles to the skeleton. Like a stretched rubber band, these tendons represent stored energy, so there is power for a rebound. Therefore, in addition to the contraction of muscle, there is a rebound of an elastic component in propulsion. This occurs in the long legs of kangaroos. The elasticity rebound provides power without much expenditure of energy. Even the basic units of muscle fibers, the sarcomeres, have their own built-in elastic proteins. However, the master of acceleration and jumping is the flea, which stores energy in an amazingly elastic material called reselin. When compressed, reselin has an enormous potential for rebound. The flea's leg, cocked and ready to leap, triggers a catch that allows reselin to expand, and this expansion, followed by the contraction of fast-twitch extensor muscles, launches the flea into prodigious jumps, more than 100 times its body length.

Using the Rebound Elastic rebound plays a major role in movement, particularly in hopping, long-legged animals like the kangaroo. Muscles are attached to the skeleton by stretchy tendons. When stretched, these elastic proteins store energy and tend to return to their unstretched state.

Light Sensors

The ability to detect light first evolved in a primitive bacteria. Within this bacteria were specialized pigment molecules capable of absorbing specific wavelengths of visible electromagnetic radiation. Once absorbed, this energy

Predator and Prey Both predator (the cat) and prey (the locust) depend on visual sensing for survival. The cat focuses on its target as it makes its slow, stealthy approach, but the locust's eyes have motion detectors which, when activated, initiate a complex set of motor acts geared toward quick escape.

evoked molecular changes and activated movement either away from or toward the light source, a phenomenon known as phototaxis. About a billion years ago, some protists developed concentrations of light-sensitive pigments called eyespots. True eyes, however, didn't develop until more complex multicellular organisms appeared about 400 million years later. These first eyes were composed of a mosaic of specialized light-sensing units called ommatidia. Each of these units was equipped with lenses and a number of photosensitive pigments. Organized into arrays of thousands of individual ommatidia, they formed structures called compound eyes. Each ommatidium was only a sensor that could respond to minuscule amounts of light energy and produce a bioelectric signal. The information provided by the

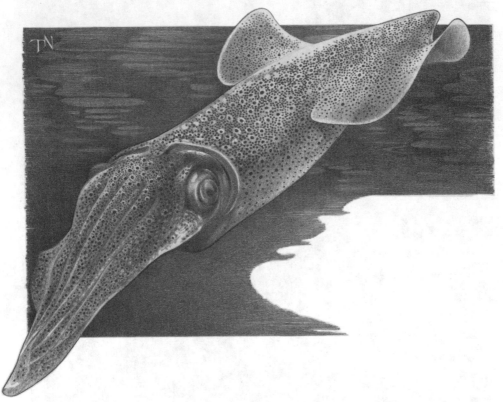

The Eyes Have It Most animals without backbones—invertebrates—have either compound eyes made up of many light sensors, or small, camera-like eyes called ocelli, or both. But one group of invertebrates, the cephalopods (squid, octopus, nautilus), has globular, camera-like eyes, whose optics—a pupil with variable diameter and a retina backed with photosensitive cells—are very similar to vertebrate eyes. However, squid and vertebrate eyes evolved independently of each other, and the developmental programs for these two eye types are very different.

Seeing in the Dark   The bush baby is a small primate endowed with excellent hand eye coordination. It can use this coordination at night as well as during the day because its large eyes can function over a wide range of light intensities. In the dark, as its eyes adapt to low levels of ambient light, the pupils dilate, the cones lengthen, the rods shorten, and pigment granules that normally shield the dim-light-sensitive rods are withdrawn. Thus equipped, the bush babies can find their prey in almost total darkness.

sensor was then integrated by neural circuits in the brain, giving the animal a visual image of its surroundings. With this equipment, the eye senses, but only the brain sees or perceives.

The compound eyes of arthropods, in particular those of insects and crustacea, are like phased arrays of radar, used to detect many objects simultaneously. Some compound eyes developed to detect patterns, sizes, and shapes, while some excelled at detecting motion with great accuracy. Others specialized in detecting color differences, and some could even detect specific wavelengths of ultraviolet light or respond to polarized light. However, a few invertebrate animals developed a different kind of eye—an eye very much like those of the vertebrates.

Squids, octupuses, and nautiloids evolved a complex eye very much like ours. It had a double set of optics that bent and focused light, a cornea, and an adjustable lens. The focused light was then projected onto the retina, which contained not only photosensitive sensors but also collections of nerve cells that started processing the incoming information. There was an adjustable opening that could regulate the amount of light entering the system as well. Some of these eyes were enormous and provided the animal with a high-fidelity visual image of its surroundings.

Despite the anatomical similarity of squid eyes and human eyes, however, they did not develop from the same primitive ancestral eye. When structures of similar form evolve using different genetic development programs, it is referred to as parallel evolution. Apparently the development of the eye structures of invertebrate cephalapods and vertebrates is one of many classic examples of parallel evolution.

The ability of prey to visually detect an approaching threat allows it to activate a very simple escape response that will remove it from harm's way. The prey may withdraw into an armored tube, burrow into the ground, dive into an existing burrow, or move away if its escape response is faster than the attack of the hunter. Likewise, the visual predator uses its eyes to detect prey, its size, its texture, and, if it is moving, its direction. Then using a more-sophisticated neural-processing system, the predator can determine how to intercept the prey's course. Arms races, of course, always lead to countermeasures. With this in mind, it is not at all surprising that predators and prey both evolved mechanisms to prevent early visual detection.

Visual Deceptions

Many animals threatened by predation evolved visual deceits that made detection by predators less likely. Among the first of these was the development of colors and patterns that closely matched their background. Later, such camouflages became increasingly elaborate: some animals could even use both neural and hormonal signals to pigment cells in their skin, enabling them to rapidly alter their color to match those of changing backgrounds. Some preying mantids look just like the brightly colored flowers of the plants that are their hunting ground. In addition to pigment changes, a great variety of animals evolved fleshy protuberances that allowed them to blend in with their environment. The predatory Saragasso fish, lying in ambush, is virtually indistinguishable from the Saragasso weed beds that are its lair. On the other hand, some predators developed highly visible, motile fleshy appendages that acted as lures, bringing unsuspecting prey into striking range.

Prey evolved still other deception techniques to thwart visual hunters. Some insects developed to look like twigs; some evolved shapes and patterns resembling bird droppings. Some developed false eyespots that confused the

Playing Dead The common garter snake, itself a predator, is often hunted by a variety of terrestrial and aerial predators. It detects prey mainly by smell, but also has quite good vision. One of its most effective tactical behaviors is to feign death when approached by a predator. Its acting is superb as it lies motionless, mouth agape, tongue dangling out. Since many predators will only eat food that they have killed, playing dead increases the odds of staying alive.

predators about where their heads were. Any visual stimulus that buys time improves the chance of survival. Some animals developed visual displays with startle value, which when activated, could often deter a predator. For example, one frog, when threatened, puts its head down and displays a startlingly large pair of prominent eye spots. Since many predators go for the head, these large, eyelike structures are effective deterrents.

Visual deception and disinformation are also effective against predators whose eyes are specialized for detecting motion. Many prey animals, seeing a potential predator coming within range, freeze in place and remain motionless. Other animals feign death when threatened; and since some hunters will only feed on living prey, this tactic is often effective. Even more elaborate visual deceptions have evolved. Some lizards, when pursued by a predator, shed a piece of their tail which continues to wiggle. This, of course, distracts the predator from the larger potential meal.

Confusing the visual hunters takes many forms. Schools of fish, wheeling rapidly in unison, lessen the odds of a predator's striking a specific member of the school. When in flight, some herd animals, such as zebras, present a bewildering array of stripes that lessens the hunting efficiency of predators, such as lions. However, the brains of predators are equipped with a highly efficient neural network that allows them to focus their attention on a specific target moving in a confusing visual background. Throughout this book, we'll examine in considerable detail the deception and disinformation strategies of a wide variety of animals. But before leaving the role of vision in predator-prey relationships, we need to examine another tactic, one that uses high visibility instead of camouflage.

Many animals and plants are beautiful but deadly, in that either they contain toxins which may sicken or kill predators, or they are equipped with highly poisonous envenomating weapons. By advertising their toxicity in bright colors and bold patterns, they improve their species' chances of survival. Obvious colorful displays also play a major role in reproduction. Plants attract pollinators with attractive flowers, and also with chemical attractants, and many animals, particularly fish, birds, and insects, use visual displays in mating.

Nerve Nets, Brains, and Behavior

As multicelled animals appeared, the workings of their diverse tissues were synchronized by two systems: the endocrine system which produces chemical messengers, or hormones, and the nervous system. The early nervous systems were relatively simple networks whose repertoire of responses was very limited. Later, multicellular animals concentrated many of their neural circuits at their

Cooperation   Musk oxen are frequently preyed upon by wolves, whose strategy is to get the herd running and then separate their target—either an older or a younger ox— for an easy kill. The predators are well trained in cooperative hunting, but sometimes the prey also cooperate in thwarting the attack by forming a defensive circle with the young in the middle and the horns of the adults turned outward at the predators.

head ends. These early brains could process sensory inputs, integrate them, and then issue commands to effectors such as muscle cells, glands, and pigment cells. Programmed in fixed patterns, hardwired reflex responses, they produce obligatory stereotyped behaviors. With the passage of time, however, complex brains evolved—brains whose data-processing machinery could be modified by experience, in other words, brains that could learn and store memories. The demands of the big brains of predators and prey were different, as were their repertoires of responses. Many of the prey animals relied on early learning at critical time periods to develop the behaviors necessary for survival. Many of these were escape and deception behaviors. The predators had a more demanding task for their brains, and most predators, particularly the mammals, had to go through a protracted training period, during which they acquired the diverse skills needed to successfully hunt prey.

Not only did the learning process involve the acquisition of the skills of stalking and hunting; it even provided for some abstract thinking in which previously learned information could be used to solve new problems. The greatest development of such innovative practices is found in humans. Unlike lower animals, humans can devise strategies and tactics of disinformation, deception, threat, and stealth beyond those produced by primary evolutionary processes, thereby making the human the ultimate predator.

Defensive Strategies

Evolution modified its earlier forms of life in diverse directions. One adaptation was to produce giants; this is particularly true of the vertebrates. There were amphibians of crocodile proportions, huge manta rays, and sharks, some of which persist in the present oceans, and during the Jurassic period, there were huge vegetarian dinosaurs, weighing 90 tons and measuring 100 feet long. In the Cretaceous period, there were flying reptiles with wing spans of 40 feet. Later, during the age of mammals, many large mammals appeared. The largest animal ever to appear on the planet was an ancestor of a terrestrial mammal that returned to the water; the great blue whale is over 100 feet long and may weigh 140 tons.

Obviously, great size acts as a potential deterrent, and predators that adapted to prey on such monsters evolved to great size. Certainly the formidable tyrant lizards of the late Cretaceous period were the most powerful and largest Earth-borne predators. So, in some cases, natural selection promoted giantism. The evolution of giant size as a deterrent to predation had a price, and a survey of giants through the ages shows that ultimately this strategy proved to be disadvantageous. Being big also means being slow and also very obvious. Giants required enormous amounts of nutrients and thus could deplete their environments. Also,

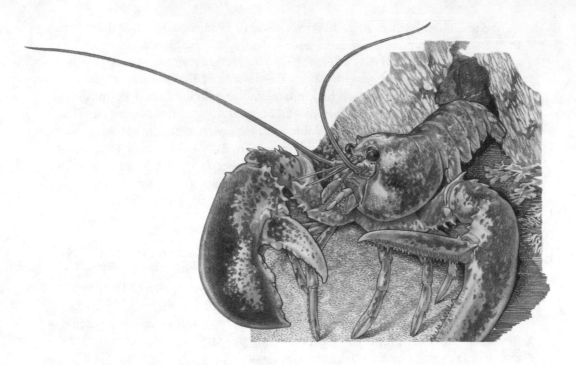

Tough Armor and Tougher Weapons   The American lobster is not only equipped with a remarkable array of sensors, but also has a dual weapons system: one of its claws is a cutting weapon; the other, a crusher. Of course, the lobster is also very well armored, and should a predator get ahold of one of its legs, it can shed that leg and make its escape. Eventually, it will regenerate the lost limb.

being highly specialized, they could not adapt well to environmental changes. This being the case, giants were often evolutionary dead ends. Being small or of moderate size seems to be the best survival strategy.

In response to the pressures of predation, prey evolved a variety of defenses to deter predators. One of these defenses was the evolution of protective armor. This strategy appeared very early in life: bacteria were protected by nonliving cell walls; protozoa built armored skeletons of silicon, calcium salts, and modified glycoproteins. The arthropods developed the most successful armor in their external skeletons, a lightweight, very strong material called chitin. But living within an armored chamber has its price. In order to grow, arthropods must molt, or shed their old armor, and then grow new, larger armor. As this molting process leaves arthropods vulnerable to predation, most seek some sort of protective shelter until their new armor hardens. Five hundred million years ago, mollusks, clams, snails, scal-

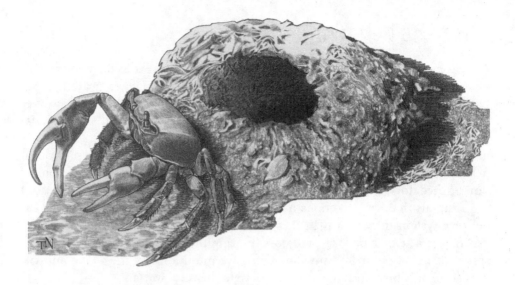

The Land Crab's Maginot Line   In order to survive attack, the land crab has an accurate early-warning system, strong defensive weapons, and an external coat of armor. However, when threatened by a predator, its best defense is to escape into its burrow.

lops, oysters, and limpets evolved very tough, rock-hard, calcareous shells into which they could retreat when threatened.

However, it was the vertebrates, particularly the reptiles, that developed the most advanced armor. The crocodile's protective armor of bony plates, imbedded in a very tough skin, makes large, adult crocodiles almost immune to predation. During the age of dinosaurs, a myriad of heavily armored herbivores evolved. In addition to armor, some of them had defensive weapons—horns, spiked tails, and rock-hard bludgeons attached to muscular tails—that could wound or kill predators. These were all eliminated during the great extinction of 65 million years ago. Today, the only armored reptiles that survived the great extinction, the crocodiles and turtles, still benefit from their armor. In the case of some of the large land tortoises, the effectiveness of their armor is attested to by their life spans of over 100 years.

Another strategy that has proven to be an effective countermeasure to predation is a burrowing mode of life. Burrowers can create both simple and elaborate safe havens by excavating tunnels into rock, wood, and soil. Provided with such sanctuaries, these animals can quickly escape to relative safety. Nonetheless, in an arms race, each successful defense brings about the evolution of predator countermeasures that can overcome it.

Chemical Defenses

A great many bacteria, protists, plants, fungi, and animals evolved a vast arsenal of chemical weapons, toxins, venoms, and deterrents that have proven to be remarkably successful. Toxins are poisonous substances, incorporated into the tissues of the organism that render it unpalatable, irritating, foul smelling, or even lethal to predators. Some of these poisons are fast acting, others slow acting. While the tissue poison might not save the organism being eaten, it does help the survival of the prey species, in two ways: the toxins reduce the population of predators, and some predators learn to recognize toxic organisms.

Animals that are venomous have a variety of stored poisons along with a delivery system through which they can inject, spray, spit, or shoot out these substances at either prey or predators hunting them. Other animals can eject sprays of irritating or foul-smelling defensive munitions. In the whole animal kingdom, only birds lack the capacity to wage chemical warfare.

In the following chapters, we will examine in detail the arms race between predators and prey.

Chapter 2

The War at Sea

Over 70 percent of Earth's surface is covered with liquid water, about 350 million cubic miles of it, with a greatest depth of 35,800 feet. The largest bodies of water, the oceans, teem with living creatures, most of which are found at depths of 1,000 feet (approximately the limit of light's penetration into clear water) or less. However, multitudes of organisms have adapted to every particular ecological niche, ranging from the brightly lit water's surface onto and into the ocean's floor, and still deeper, into the high-pressure, cold Stygian darkness of the oceanic abyss. To hunt in this tridimensional, aquatic environment, organisms evolved a number of highly specialized systems: sensory systems to detect and home in on targets, communication systems to allow collaborative efforts in both hunting and evasion, navigational systems to plot courses from place to place, propulsion systems to provide sufficient speed to overtake a prey or maneuver out of harm's way, and an extraordinary array of specialized weapon systems and defense systems.

Sensory Systems for Aquatic Environments

Since sound and other vibrations can be sensed much better in a dense medium like water than a thin one like air, it is not at all surprising that a number of marine organisms evolved both passive and active sound-location systems. With a passive vibrational detector system, a predator can home in on vibrations caused by an object moving in the water. An active sound-locating system (sonar) uses self-generated bursts of sound and their reflected echoes to locate unseen prey and obstacles. Neither of these tasks is easy, because the oceans are filled with noise, and sound waves propagated through water not only decrease over distance but also may be bent by water layers of varying densities. Yet, the abilities of echo-locating marine mammals have been fine tuned by natural selection over millions of years. Not only can they generate sub-sonic and sonic pulses, but they can also detect the frequency intensity, band-width, pulse shape, and directional characteristics of the reflected echoes. All this information enables them to separate out background noise and get an extremely accurate image of both the size and velocity of the target, as well as its direction.

Water is also a good conductor of electricity; so naturally, a number of aquatic animals developed the capacity to detect very slight electric currents. Moreover, they evolved the capability of generating their own electric field and then using distortions of that field to detect underwater prey, avoid obstacles, and communicate with others of their own species. Some freshwater and marine organisms even evolved a weapon system that uses very strong electric discharges to stun or kill prey or to deter a potential predator.

Navigation at Sea

When humans first went to sea, they stayed close to the coastlines. These areas provided them with visual landmarks that secured their sense of location. This type of navigation is known as "pilotage." Later, humans learned to use the geo-magnetic compass; the Sun and the stars; radio waves; and gyroscopic, stabi-lized inertial guidance systems to plot their course across unfamiliar seas from embarkation point to destination and back.

Navigation involves plotting a direct course between two points, from *A* to *B,* using spatial memory and a variety of reference points. Spatial memories learned by humans are recorded and accessible in the form of maps and charts. But do we and other animals also have the capacity to record spatial memories in our brains? And can these memories be part of a genetic endowment? The answer seems to be yes.

Many animals besides humans have a sense of location. That is, they know where they are, and if they go someplace else, they have a sense of where they came from and can return to their home port. A number of studies of migratory species show that genetically transmitted spatial mem-ory of landmarks is involved in some of these rather remarkable journeys. Sometimes the innate program requires the organism to learn to recognize landmarks automatically during some early stage of its postnatal develop-ment. To date we know of animals that use vision, smells, sounds, magnetic lines of force, the Sun, the stars, polarized light, ultraviolet light, water cur-rents, temperature differences, pressure waves, and even electric field varia-tions to orient themselves. These extraordinary sensory systems provide landmarks.

Following Its Nose The adult green turtle, shown here, migrates from the coast of Brazil to Ascension Island, the place of its birth. The young turtles mature in Ascension and learn the distinctive odors in the water there before following the ocean currents that bring them to Brazil. Years later, when sexually mature, they swim against the current and home in on the memories of the distinctive chemical signatures in Ascension's waters.

Some animals and humans also have an innate orientation program, which differs from real navigation. Orientation is the tendency to go in a given direction regardless of the starting point. Some animals, such as night-migrating birds, will always orient themselves in the direction of the North Star and the pattern of constellations that surround it. They learn to do this unconsciously, and when their built-in biological clocks activate the orienting behavior, they head off on that path.

A number of fish, marine mammals, and sea turtles are known to perform remarkable migrations covering thousands of miles. Each of these long-range missions requires considerable navigational skills, skills developed long before humans became accomplished seafarers. Sea turtles and salmon are particularly worthy of note.

The green turtles, *Chelonia mydas*, breed on a tiny speck of volcanic island in the middle of the South Atlantic Ocean. Ascension Island is only 6 miles wide and 7 miles long and is 1,500 miles from these turtles' feeding grounds off the coast of Brazil. Every two or three years, sexually mature adult turtles are automatically triggered to orient themselves back toward Ascension Island and start swimming "up current" across the vast emptiness of the Atlantic. The trigger is thought to be an environmental signal that causes hormones to be released. These hormones in turn activate the migratory drive circuits in the turtles' brains.

Several theories have been put forth to explain how these turtles navigate. Some have suggested that they use the Sun or polarized light as a guidance system; others have suggested they use celestial clues. Recently, others have suggested magnetic-compass-type mechanisms. All these types of navigation have been demonstrated in other animals. But in turtles another system seems to be used, a system yet to be mastered by human navigators: chemical sensing.

Green turtles lay their eggs on the sandy beaches of Ascension Island and then return to the sea, leaving the Sun's heat to incubate the eggs. Long after the adults have departed, the eggs hatch. The hatchlings consistently and unerringly move toward the ocean, even though they can't see the water. Apparently, they have an innate program that makes them move toward the bright light of the beach and not toward the darker, inland landscape, where they would most certainly die.

Once in the ocean, the young spend some time exploring the waters around their hatching site. During this time, they imprint on the characteristic chemical signature of the waters around Ascension Island. Imprinting is a special kind of learning that occurs at specific critical times in an animal's life. The end result of this imprinting is the establishment of some well-fixed, long-lasting circuits in the turtles' brains. The young turtles then migrate "down current" to their feeding grounds off the coast of Brazil. Later, when they are adults, they take in their turn the long "up current," swim to the chemical signature of Ascension's waters.

Researchers believe that this process of chemical navigation evolved over millions of years when Africa and South America were joined. As the continental plates began to drift apart, ancestral green turtles traveled only short distances from what is now the Brazilian coast to the nearby volcanic island. Over the span of 100 million years of geologic time and continued drift, the journey got longer and longer, but the innate chemical-sensing system continued to work.

The life history of salmon apparently involves a similar chemical-sensing form of navigation. Adult salmon are wide-ranging ocean dwellers for most of their lives, but, when sexually mature, they return to the same streams in which they themselves were spawned. There is good evidence that young salmon, like green turtles, imprint on the chemical signatures of the streams of origin and, on their return "up current," follow chemical clues.

Navigation by Salmon   Salmon hatch and develop in freshwater streams, each having its own distinctive chemical imprint. When big enough, the salmon swim downstream into the oceans, where they spend most of their adult life. When ready to mate, an innate drive is activated, and they head back to the waters that were their roots. A high percentage even return to the same stream in which they hatched. Somehow they retain a long-term memory circuit that recognizes their early chemical environment.

Advanced Propulsion Systems in Aquatic Animals

In late 1944, the Allied air forces in Europe suddenly found themselves facing a formidable new opponent, the German Messerschmidt 262. This fighter could fly 250 miles per hour faster than the B-17 bombers and 100 miles an hour faster than the best Allied fighters because they were powered not by the usual reciprocating engines but by an engine that pushed them forward with a jet of hot gases. The jet age was upon us.

Aquatic animals, particularly the ocean-dwelling cephalopods, had been using this means of propulsion for hundreds of millions of years. The use of water jets for propulsion first evolved more than 500 million years ago in a group of mollusks called the cephalapoda (head-footed ones). Today, their descendants—squid, octopuses, nautiloids, and cuttlefish—are found in all the oceans of the world. Ancestors of all the cephalapoda bore shells, composed of partitioned spirals, which are preserved to us as fossils of great geometric beauty. Once there were at least 10,000 related species, but their numbers gradually diminished and, after the great extinction, few survived. Of the shelled forms, only the chambered nautilus and King Nautilus are found today.

Modern nautiloids are jet-propelled, solitary predators. During the day they rest on the deep, cold ocean bottoms, where there is relatively little oxygen. To survive in this niche, they lower their metabolism and use oxygen trapped in the buoyancy chambers of their shells as "scuba" tanks. (Evidently, the ancient cephalopods were innovators not only in the use of jet propulsion but also in the use of self-contained diving gear.) Each night modern nautiloids use their buoyancy and jet propulsion to make long, vertical migrations to the surface to feed. They have a good sense of smell to locate prey and carrion, and an array of 90 tentacles to grasp and hold the prey when found. One of nature's enigmas, they are literally living fossils of which we know relatively little; however, their cousins without shells, squid and octopuses, have been extensively studied.

Octopuses are intelligent ambush hunters that lurk among the crevices and caves of rock- or coral-lined shallow seas. Masters of camouflage, they are capable of rapidly expanding the pigment-containing cells of their skin to match their surroundings. Unlike those of their sleek, streamlined cousins, the squid, octopuses' bodies are bulbous, highly flexible masses with eight suction-cup-laden arms. The octopus uses its arms to pull itself around the seabed, but it too has a jet propulsion system. A superb visual hunter, it can quickly pounce on a prey that comes within range. With the suckers on its tentacles, it latches onto its prey, usually a crustacean, and pulls the prey toward its mouth cavity. Once it has grasped the prey in its sharp, horny beak, the octopus administers the coup de grâce by injecting a potent nerve

Jet Propulsion with a Smoke Screen   All octopuses are marine mollusks armed with eight prehensile tentacles, each of which is equipped with suckers of various sizes. Being shell-less and soft-bodied, they can squeeze through narrow openings. These cephalopods, cousins to squid, cuttlefish, and nautiloids, have keen vision and can use their water-jet propulsion systems for rapid escape. As they shoot backwards, they emit blobs of inky mucous, which distracts their predators. The ink also deadens the sense of smell that their main predator, the moray eel, uses to locate octopuses.

poison produced by its salivary glands. The poison quickly paralyzes and kills the prey. One small species of octopus found off the Australian coast, *Octopus maculosus,* known as the "blue-ringed octopus," has been responsible for a number of human fatalities and was featured in one of the James Bond films, *Octopussy.*

Though well armed with potent offensive weapons, the octopus, when confronted by a predator, flees rather than fights. Like the squid, it compresses its body and jets away backward, tentacles trailing behind while it emits a covering "smoke screen" in the form of a blob of inky mucous.

Squid are exceptional swimming machines also propelled by water jets. They suck water into their cylindrical, tapering, muscular bodies via a pair of valvelike vents on either side of their head. When they contract these muscles, the water in the cylinder is compressed and vented via a funnel-like nozzle. The nozzle is under the squid's conscious control and can be pointed fore or aft on command, permitting the jet to push the squid forward or backward through the sea. Usually squid swim forward at a leisurely pace, about 1 foot per second when migrating. At this pace, the Japanese squid, *Todarades pacificus,* covers over 2,000 kilometers in less than three months. However, when escaping from predators, squid are able to generate a sudden acceleration backward that can carry them out of harm's way at the speed of 25 body lengths per second. Often as they burst away from an approaching predator, they release a blob of dark pigmented mucous (or ink) that confuses the hunter and buys time for escape.

The rapid burst of speed of the squid depends in part on the organization of the fast-twitch muscle fibers that compose 80 percent of its container, or mantle, and in part on elastic collagen fibers, which are stretched when the mantle is filled with water. Working like a stretched rubber band, this elastic rebound speeds the contraction. In any escape response, speed is of the essence. In this case it is provided by giant nerve cells whose axons' enormous diameter allows for rapid conduction of signals to the muscles.

Recently it has been discovered that squid have yet another time-saving mechanism. When nerves talk to the muscles, they release information-carrying molecules, or neurotransmitters, at junctions called synapses. In most animals, this process takes time (which is why it is called synaptic delay), but in squid, the synaptic transmission is very fast. With this combination of rapid processing and conveyance of the escape command and rapid contraction of the mantle, squid have a significant survival advantage.

Squid are prodigious hunters. Prey captured by the squid's suction-cup-armed tentacles is brought to its mouth, where a powerful parrot-like beak grasps and kills the victim and a rasping, muscular tongue tears digestible-sized pieces off the prey. The most mysterious of the squid are the giants of the genus *Architeuthis.* These monsters may weigh up to 1,000 pounds and measure 60 feet in length. They have huge heads dominated by a pair of very large eyes. The head connects via a short neck to the muscular mantle. At the front of the head is a crown from which extend eight 9-foot-long muscular arms and two 30-foot-long tentacles. The arms and the paddle-like expansions at the ends of the tentacles have rows of suckers and adhesive knobs for grasping their prey. The largest of the suckers are set on muscular stalks and have diameters of 2 inches or more. Each suction cup is ringed with sharp teeth made of chitin, the same material found in lobster shells.

We don't know much of these giants that range from the surface to the ocean depths. We have never captured one alive, but we know they have a wide range. There must be a lot of them around, judging from the frequency with which their beaks are found in the stomachs of sperm whales. According to recent estimates, there are over 100,000 sperm whales in today's oceans, and if

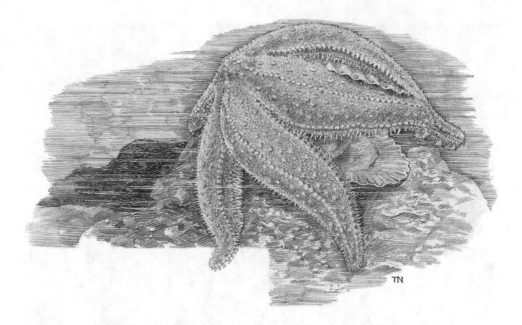

Slow-Moving but Deadly   Although slow-moving, the starfish finds plenty of food by hunting stationary prey, such as the clam, shown here being wrapped by the starfish's arms. Although clams are armored and can lock their shells shut, they have a chink in their armor: the thin, soft tissue between their two shells. The starfish extrudes its stomach and digests its way into the clam, then sucks in its ingested meal.

each whale feeds on a dozen giant squid per year, there must be millions of them. Despite this relatively large population, we have yet to see one alive in its oceanic habitat. Nevertheless, somehow they have captured the human imagination. Recently, Peter Benchley, the author of *Jaws,* has found a new book topic, a giant squid that he calls "The Beast." A century ago Jules Verne, author of *20,000 Leagues Under the Sea,* also featured a giant squid attacking his mythical submarine, the *Nautilus.* Other books and movies have featured another jet-propelled cephalopod, the giant octopus.

Jet-Propelled Aquatic Animals

Scallops, like most bivalved (two-shelled) mollusks, are filter feeders that don't pursue prey. However, they are hunted by a variety of predators, including man. Their primary defense is a protective armor, namely their hard, closable, calcareous shells. Unlike other bivalves that also deter predators by burrowing into the sand, scallops remain exposed on the surface of the seabed, counting

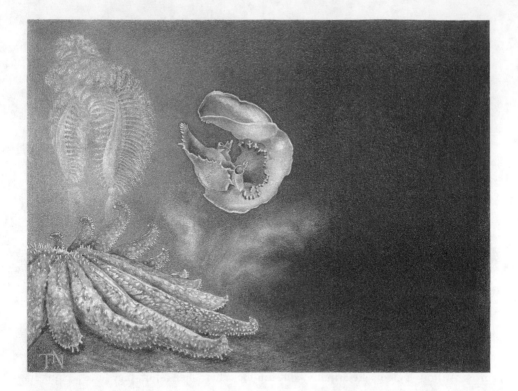

A Jet-Propelled Escape   Several mollusks can detect the odor of a prowling starfish and escape by using a jet of water generated by contracting their muscles. These mollusks' escape speeds are faster than the slow-moving starfish and if such a mollusk detects the predator, it usually gets away.

on their armor to protect them. However, there are two very slow predators that can penetrate bivalves' tough shells. One is a snail equipped with a specialized rasping organ that drills through the shell. The success of these gastropod mollusks, called oyster drills, is attested to by the large number of shells found on the beach that have a single neat, round hole in them. Starfish, the other bivalve hunters, are even more effective than oyster drills. They engulf a bivalve in a lethal embrace, holding onto the shell's surface by the suction of hundreds of tube feet. We originally thought that the starfish's pull eventually fatigued the bivalve's adductor muscle, causing the shells to open wide enough for the starfish to extrude its stomach into the opening and digest the bivalve's innards. This, however, is not the way it works; adductor muscles don't tire out. What happens is that the attached starfish extrudes its stomach and digests its way in at the margins of the closed shell. It insinuates its stomach through this very thin aperture and then kills and eats the clam.

Scallops under predation pressure by starfish adapted by evolving an escape mechanism that got them away from the slow-moving hunter. Scallops have a whole row of metallic blue eyes and possibly a good sense of smell with which they can detect an approaching predator. A few drops of ground-up starfish put in the water near a scallop will cause it to lurch upward suddenly and swim forward by clasping its shells together. Scallops can swim about 2 miles per hour, but they swim only occasionally and cover a distance of only a few yards. To achieve this, the scallop shuts its hinged shell, compressing a large volume of water which is then squirted out of two channels on either side of the hinge, thus jetting the scallop forward. When the scallop lands and relaxes the adductor muscle, the hinge automatically pops open due to elastic rebound of a hinge protein called abductin. (The resilience of this rubber-like protein is comparable to reselin, which powers the leap of the flea.) In this way, almost 90 percent of the energy stored during closure of the shells is released to open the shell with minimum energy expenditure. Though scallop swimming produces a rather crude, seemingly clumsy movement, it does the job of taking the scallop out of danger.

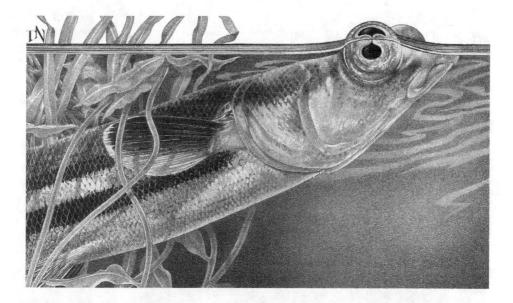

Up Periscope: The Four-Eyed Fish   *Anableps dowi* is a small, tropical, freshwater fish that frequents the water's surface. It has evolved four eyes—one pair for searching light coming from the air, and the other for detecting objects in the water. Thus, it can survey both its aquatic and aerial surroundings for friends, foes, and prey. These unusual eyes have two pupils that regulate the amount of light coming from above or below.

Many species of fish use muscle power to project themselves out of the water in order to capture airborne prey. Among the more familiar leapers are salmon, trout, and bass. But the monkey fish of the Amazon basin, the Arowana, is the most spectacular of the leapers. It turns itself into an optically aimed, submarine-launched ballistic missile. This requires some rather elaborate aiming corrections to compensate for the refraction of light at the air-water interface.

Another leaping aquatic hunter has evolved an optical system that doesn't need to make corrections for refraction: it uses a periscope. *Anableps dowi*, a small freshwater predator of Central America, has evolved eyes with two optical systems. The upper half of the eye protrudes above the surface while the lower half scans below the surface. With such an optical system present, the fish remains concealed from flying prey before launching itself; at the same time it can look down in search of predators.

Leaping uses a great deal of energy. In terms of benefit to the hunter, this strategy is of questionable worth, since the prey is usually a small insect of low caloric content and the energy expended in the launching process exceeds the energy obtained in the meal. Why this strategy has persevered so widely and for so long is not understood.

Weapons—Jaws 1

The fore limbs and hind limbs of vertebrates modified for propulsion and maneuver when they moved back into water; however, these modifications

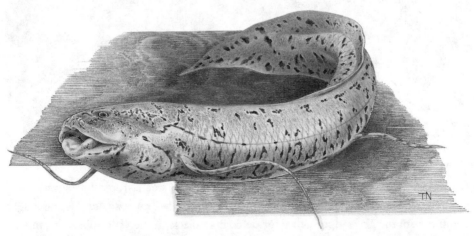

A Voracious Freshwater Predator Although the African lungfish lacks real teeth, it has sharp, overlapping, bony plates that act as shears and can rapidly shred prey. Thus it is the top predator in some African rivers.

deprived them of a major weapon system, claws and talons, to capture prey. Their tooth-lined mouths became the primary weapon system of most of the aquatic vertebrates. Of course, teeth and jaws come in many varieties. This and the following three sections will examine the weapon jaws in detail.

Seven hundred million years ago, when the basic genetic blueprint for all advanced multicellular animals was laid down, one program included a plan for a food-obtaining apparatus at the head end. Over the course of time and under a great variety of selection pressures, many different structures evolved. The usual pattern involved some form of biting jaws to grasp, slash, stab, cut, or otherwise obtain needed nutrients. Humans are particularly fascinated by toothed jaws. But there are many ways to bite. So before looking at the familiar big biters, let us examine some of the less familiar mouths of the animal world.

Protopteurus annectus, the African lungfish, is an obligate air breather that is called fish. In reality, though, it belongs to a class of its own. African lungfish are fat-bodied predators that can be as long as 6 feet. They have adapted to periodic droughts by creating mucous-lined cocoons in the mud where they sit until the rains come. Sometimes they can stay in this state of hibernation for

The Compleat Angler   The ugly bottom-dwelling monkfish sits on the bottom, its mouth agape, waiting for prey to come into range. With the skill of Izaac Walton, it uses a fleshy lure at the end of a spine to entice prey within range of its gaping maw. Once the prey is close enough, it suddenly opens its huge mouth, creating a strong current that sucks the prey into its lethal jaws.

four years; when they emerge, they are much thinner and very hungry. Ambush hunters with catholic tastes in food, they will eat anything they can catch. In some African lakes they are the top carnivores. Detecting prey with their chemoreceptors and lateral line vibration receptors, they accelerate with an incredible burst of speed and then snap up their target with their toothless jaws. However, their jaws are composed of sharp-edged, overlapping, enamel-coated bony plates that act as shears. The effect is that of an aquatic Cuisinart. Lungfish are also very mean, as anyone who had gone on a four-year diet would be.

Luring prey into their jaws' range is a widespread adaptation seen in a variety of motionless ambush hunters. Such a strategy conserves energy but requires the evolution of some form of enticement to attract would-be predators into striking distance. In marine environments, more than 200 fish of the order *Lohiiformes,* the anglerfish, have modified the first spine of their dorsal fin by moving it forward onto the snout. The spine is flexible, and at its tip is a fleshy appendage that acts as a lure. In some anglers of the dark ocean depths, the lures contain light-emitting bacteria. Shallow-water forms, in contrast, use simple lures that look like edible morsels. The complete angler of the genus *Antennarius* is found in the coral reefs of the Indopacific. Not only is it concealed both by its color pattern and by bulbous fleshy structures that mimic its surroundings, but its lure looks very much like a small fish, replete with fins, eyespots, and pigmented bands.

In northern waters, the common anglerfish, or monkfish, uses its fleshy lure as an appetizing decoy. With the skill of an experienced fly fisherman, the monkfish moves its lure around, just in front of its partially agape, huge, tooth-filled mouth. These teeth, hundreds of them, are hinged so that when a prey is ingested, they fold back, because they aren't for biting. When the monkfish suddenly opens its cavernous maw, the influx of water creates a current so strong that it sweeps the prey in; then, the jaws close, and the teeth become erect, trapping the victim. Monkfish can get quite large, up to 6 feet, but despite their size, they are masters of concealment on the seafloor. Their inconspicuousness is due in part to their mottled camouflage but also to fleshy seaweed-like excrescences all over their bodies, particularly at the margins. Their heads are so grotesquely ugly that fish vendors remove them for fear of discouraging customers. That being said, monkfish is very tasty.

Some mouths lack true jaws that open and shut, like the primitive lampreys and hagfish, two of the major predators of freshwater and marine fish. These slimy-skinned, eel-shaped ancestors of the vertebrates lack vertebral columns; support is provided by an elastic rod that runs along their back and is attached to bundles of cartilage. Their circular mouths lined with rows of curved teeth act as suckers, and their whole mouth cavity is fringed with fleshly flanges that allow the sucker to seal onto the skin of their victim, a fish. Once locked onto its target, the lamprey uses its powerfully muscled toothed tongue to rasp its way into its victim's flesh. While the sucking and rasping weaponry works well, it does not do as well as the powerful toothed jaws of the notorious piranha of the Amazon basin.

Tales abound about this small South American and Central American fish endowed with jaws lined with a single row of triangular sharp-edged teeth; indeed, the name *piranha* means "toothed fish." When the piranha's powerful jaws close, the teeth of the upper and lower jaws intermesh like a bear trap, slamming shut to chop off bite-sized pieces of flesh. Although the notorious red piranha is relatively small, most other types of piranha are about 12 inches long. They travel in hunting groups called shoals, which may contain thousands of fish. There are authenticated reports of a shoal stripping a cow to a defleshed skeleton in a matter of minutes, as the shoal seems to go into a blood-induced feeding frenzy. Naturally, such gruesome stories are the stuff that movies are made of, and next to sharks, piranhas are Hollywood's favorite.

But what is the reality? My colleagues who have spent lots of time in the Amazon and Orinoco basins' waters which are full of piranhas have never been bothered. As a matter of fact, they fish for piranha, which they say are very tasty. They attribute the tales of piranha's feeding frenzies to incidents of fish trapped in large puddles after river flood tides subsided. In these situations, the frantic, trapped, starving fish exhibit bizarre behavior. Clearly, piranhas are great little aquatic biters that capture the human imagination. But the real goblins of our bad dreams are the big biters—the sharks.

Jaws 2—Sharks

Sharks are humankind's worst nightmare. As terrors of the deep, they have been cast as villains in countless books and movies. There is some basis for such fear. In the waning days of World War II, the *U.S.S. Indianapolis* was torpedoed by the Japanese submarine I-58 and sank quickly, leaving over 800 survivors in the water. The first circling shark appeared after a few hours. With each passing day more and more deep-sea sharks appeared, and hundreds of sailors ended up in sharks' bellies. Other marine catastrophes recorded similar shark attacks. But the reality is that human deaths from shark attacks are quite rare—only about 50 deaths per year, far fewer than those caused by bee stings or eating toxic shellfish.

The public image of the shark as a toothy, torpedo-shaped, fast-swimming, relentless predator is based only on those few large species known as "man-eaters." The acknowledged top gun of these is the great white shark of

Hollywood fame, *Carcharadon carchanias* (Greek "ragged toothed"). When I was a boy I used to collect fossils from the crumbling cliffs along the Chesapeake Bay; the most abundant of them were sharks' teeth. Some were huge, 5-inch-long triangular blades with serrated edges, and I wondered about the animal that had left those dental remains. Later, when visiting the Smithsonian Museum, I saw a plastic restoration of the jaws that bore those teeth, the jaws of *Carcharodon megalodon,* an ancient ancestor of today's white sharks. From the jaws, it was estimated that *C. megalodon* was about 100 feet long; however, it was later shown that the initial construct was too large and *C. megalodon* was a mere 40 or 50 feet long. An accurate reconstruction of its jaws at the American Museum of Natural History in New York shows they could gape wide enough to hold a Volkswagen Beetle.

Sharks originated 350 million years ago in the Devonian period, before fish filled the oceans. They are vertebrates whose skulls and skeletal structures are made of hardened cartilage; consequently, they left little fossil record except for their calcium-containing teeth. Judging from the size, shape, and numbers of those fossils, the ancient seas teemed with thousands of species of sharks, as well as rays and chimeras (both cousins to the shark), most of them small. Shark jaws may hold up to twenty rows of teeth, and as the shark grows, teeth are shed and replaced by new developing teeth. In a lifetime, sharks may produce over 20,000 teeth. Some, like the bug-eyed, deep-dwelling, slender crocodile shark, have long, narrow-cusped teeth. Deeper yet at 9,000 feet are the ugly, flat-snouted, flabby goblin sharks, which have slender, needle-like teeth. Larger sharks, like the white tip, have triangular teeth and can bite down with a pressure of 22 tons per square inch, a force 300 times greater than a human bite.

Surprisingly, shark brains are much larger than those of fish even though sharks appeared earlier in evolutionary history. A disproportionate volume of their brain is dedicated to processing information about smells; in fact, some people call sharks "swimming noses of the sea." Black tip sharks can detect one part of extract from grouper in 10 billion parts of seawater, and requiem sharks

Giants of the Prehistoric Seas (facing page)   The warm, slightly salty seas of a hundred million years ago teemed with both invertebrate and vertebrate animals, both predators and prey. Some of the giants of the ancient seas shown here are *Ichthyosaurs* (top), *Elasmosaurus*, a long-necked plesiosaur (left), and the ferocious, toothed *Tylosaurs*, a mosasaur or marine lizard. Related to the dinosaurs, they all became extinct 65 million years ago. There were also giant invertebrates like the 30-foot-long squid (right), whose relatives still inhabit the modern seas. In addition, there were giant cartilaginous fish like *Chacharadon megalodon* (bottom), an ancient 40-foot-long shark whose enormous, tooth-lined jaws could swallow a Volkswagen Beetle. Although extinct, this huge shark is thought to be a distant ancestor of today's great white shark.

one part of tuna juice in 25 million parts of water. It has been amply demonstrated that sharks can detect blood in the water at a long range; sometimes, albeit rarely, blood in the water triggers the animals to go into a feeding frenzy.

Although they are primarily olfactory hunters, they also use a remarkable array of other sensors in hunting. This collection of sensors has kept them at the top of the oceanic food chain for 200 million years. Their vision, once called into question, is as remarkable as their sense of smell. Deep-sea species have large eyes with lots of rods and reflective retinas, while those who hunt near the surface have keen vision provided by retinas richer in cones than the human eye. Although earless, they can detect sound vibrations of very low frequency, well below the level detection limit of human hearing, by means of fluid-filled pores along their sides. However, the sharks most sensitive detector system is housed in jelly-filled ducts in the shark's snout; it can detect astonishingly small electric currents.

Some rays and some fish such as those found in murky waters in Africa and South America also have electroreceptors. None, however, are as sensitive as those of a shark, whose electrosensors can detect the tiny direct-current field generated by any living things swimming in water. Within range of a direct current up to 8 hertz, sharks can detect a hundred millionth of a volt per centimeter. Some experimental findings suggest that this great sensitivity to voltage may be the signal that initiates attack. Recently, AT&T, using a new submarine lightweight cable experienced a sudden short circuit. On examining the cable, the company's technicians found that it had been bitten by sharks. Apparently, the extraordinarily weak electric fields generated by the submerged cable provoked them to attack.

Sharks use this great mass of sensory information in their hunt for food. Since the advent of the four *Jaws* movies, we have gained access to some extraordinary videotapes of great white sharks attacking seals and sea lions off the California coast. These videotapes and other research give us a truer view of shark behavior than was fancifully concocted by the screenwriters in tinsel town.

Although the great whites are responsible for many of the recorded attacks on humans, their food preferences, as adults, run to high-calorie, high-fat, high-cholesterol meals provided by the insulating blubber of seals and sea lions. (When digested, fat yields 9 calories per gram, more than twice the calories in a gram of carbohydrate or protein.) In other words, sharks are not vicious indiscriminate hunters but hungry predators responding to their nutritional needs. Most of the particular attacks recorded in the videos were aimed at surface-swimming pinnipeds early in the day, indicating that when hunting, great whites rely heavily on their acute vision. Although they don't hunt cooperatively, as do dolphins and killer whales, they do establish hunting zone territories near the shores, where they position themselves for the attack on the pinnipeds coming ashore. These danger zones are well known to the big-brained marine mammals. They know the sharks are waiting, cruising slowly along the bottoms ready to surge upward and attack animals silhouetted against the surface. To counter this threat in the danger zones, elephant seals stay away from the surface, while sea lions swim on the surface in tight formation, frequently breeching in a cooperative system designed to confuse the predator.

Seals appear to be the most favored prey, possibly because they lack the sea lion's well-developed front flippers which can deliver a powerful slap to the shark's sensitive snout. Sharks attack the seals from below and behind, and usually bite them in their highly vascular heads. With its prey locked in its jaws, the shark descends, leaving a long trail of blood; he then bites off a big chunk of flesh and lets the exsanguinated carcass float away briefly until he returns to feed some more.

The attacks on sea lions are carried out with an explosive burst of speed that carries the shark and its seized prey out of the water. The shark pulls the mortally wounded sea lion down and then releases it after tearing out a chunk of flesh. Bleeding, the prey swims to the surface, only to be attacked again. This process is repeated again and again. Similar strategies are used in attacking other blubber-rich marine mammals, such as whales and dolphins. Since adult sharks increase their size by 5 percent per year, they need an enormous caloric intake to sustain this growth. There is debate among researchers regarding how big they can get, but it is substantiated that some males grow to more than 20 feet in length.

Sharks compete with each other for floating bloody kills. This competition takes the form of a ritualized behavior involving swimming near the surface and slapping their tails into the water. Apparently it is a contest of power, and it always ends with one competitor's leaving the arena vanquished and the victor's taking the spoils.

No animals other than humans prey on great white sharks. Dolphins, however, do fight back. There are a number of well-documented cases in which a group of dolphins have repeatedly charged at sharks, battering the shark's sides with their bony beaks. Throughout this harassing behavior by groups of dolphins, they continuously communicate with each other, bracketing the shark on either flank, butting and withdrawing before the shark can counterattack. While the dolphin lacks the shark's awesome weaponry and sensory armory, it does have two great advantages: the ability to generate and use sound, and a big brain that allows it to learn and remember.

Jaws 3—Aquatic Reptiles

During the Age of the Reptiles of the Mesozoic era, a great variety of reptiles evolved and dominated the earth for more than 100 million years. Of these, one major group, the ruling reptiles, or dinosaurs, diversified and gave rise to a large variety of now-extinct marine forms as well as the only survivors today, the birds and the crocodilians.

Crocodilians arose from a group of ancestral reptiles called thecodonts. Like the modern crocodilians, these thecodonts had sharp, pointed teeth set in bony sockets in their lower and upper jaws. Most were small with long hind legs and short forelegs. Many of the thecodonts evolved a bipedal gait, which gave them a considerable speed advantage, given their long hind legs. Some, however, reverted to a four-legged gait, retaining, nonetheless, shorter forelegs, a characteristic common to all crocodilians. Some of these reptiles also developed rows of bony plates in their skin, an armor plating retained, to varying degrees, in both ancient and modern crocodilians. The increased weight of armor encouraged some of these four-legged thecodont progeny to take to the water, where their weight was offset by buoyancy. Of these early water dwellers, some became exclusively aquatic, while others became amphibious and frequented the banks of rivers, lakes, and estuaries.

All of these aquatic predators were air breathers. Over time, they developed several adaptations permitting them to stay submerged for long periods. Those forms which hunted along the shores for land dwellers coming to drink exhibited another evolutionary strategy: the ability to submerge most of the body while leaving the air-breathing tube, ears, and eyes just above the waterline. In developing their snorkeling strategy, the early ancestors of the crocodilians located their breathing apparatus on top of their heads between their eyes. This gave them a direct, short air passageway to the lungs; however, it also deprived them of an important sensory system: because their nostrils were submerged, they could not use the sense of smell to locate prey. This design defect was corrected early in the history of the crocodilians by moving the nostrils to the front of a long snout. With this change crocodilians gained the advantage of being able to submerge almost all of their bodies, leaving only the tops of their heads exposed, while still having full use of all three sensory systems—visual, auditory, and olfactory.

Although the fossil record of the ancestral crocodilians is not complete, it is still possible to construct a reasonably good history of crocodilian evolution. The success of their evolutionary adaptations is attested to by the fact that they alone, among the ruling reptiles, survived the great extinction that occurred 65 million years ago.

Some of the early crocodilian ancestors, the mesosuchians, underwent structural adaptations to an aquatic life. The arrangement of their teeth in long, thin snouts suggests that they were primarily fish eaters that were totally adapted to the marine environment. Instead of legs, these ocean dwellers had paddle-like appendages; they lost their armor and evolved a large dorsal fin near the end of their long, muscular tails. Although some of these specialized

Armored Snorkelers   Large crocodilians are virtually immune to predation. Protected by an armor-plated skin, these powerful predators are awesome killing machines. They can detect prey by sensing odors, sound, and vibrations, and are also endowed with keen vision. Almost invisible, their approach is in stealth, but the actual attack, a high-speed lunge, brings the prey within range of their powerful, tooth-lined jaws.

dolphin-like crocodilians were successful, eventually they became extinct in the early Cretaceous period. Other mesosuchians developed adaptations similar to those of modern crocodiles.

Fortunately for researchers, the crocodilians were semiaquatic and, on their deaths, frequently became embedded in mud; as a result, they were readily fossilized. One fossilized ancestral crocodile, a giant that thrived during the Cretaceous period, the Age of the Dinosaurs, was called *Phobosuchus* (Gk *Phobus* "fear" + *souchos* "crocodile"). Recently, this now-extinct member of the subfamily Crocodylinae was renamed *Deinosuchus* (Gk *Deinos* "terrible"). It had a 6-foot-long skull, reached a maximum length of 12 meters (40 feet), and weighed several tons.

Modern-day crocodilians—crocodiles, alligators, and gharyals (or ghariyals)—are all superb, cold-blooded hunters, marvelously adapted to the role of stalking, capturing, killing, and eating a variety of prey. The long-nosed species, the gharyl (*Gavialis gangeticus*), which resembles ancient ancestral forms, is primarily a fish eater that captures its prey by means of a rapid sideways snap of its tooth-lined jaws. The advantage of the long, thin snout is that, during the quick, horizontal hunt snap, it can move through the water with much greater ease than can a broad snout.

The broad-nosed species have diets that vary widely with age and availability of prey. During their first years of life, they subsist on a diet of insects, spiders, crabs, small frogs, snails, and even aquatic insect larvae. They round up their prey by curving their bodies and snapping them up with a visually aimed sideways movement of the jaws. As they grow, their diet changes, and they begin to eat small mammals, aquatic birds, turtles, and other reptiles, including smaller members of their own species. Indeed, whenever one observes Nile crocodiles basking on the sandbars and riverbanks, one is struck by the fact that in any grouping, all individuals are about the same size. This behavior is thought to be a self-preservation tactic that prevents the large crocodiles from cannibalizing their smaller cousins.

Most stalking and hunting occurs in the water and at the water's edge. The initial approach is slow and stealthy. The crocodile is almost totally submerged; only its eyes, ears, and nostrils show above the surface. It propels itself forward by undulating its tail, and while swimming, it assumes a streamline configuration by holding its fore limbs and hind limbs closely against its body.

The approach is so slow that barely a ripple is detectable. As the hunter gets within range, it submerges completely until it is within 4.5 to 6.0 meters (15 to 20 feet) of its prey. The attack is a sudden, rapid strike propelled by a few strokes of the crocodile's powerful, muscular tail. With the enormous force of more than 200 kilograms per square centimeter (3,000 pounds per square inch), the vicelike, tooth-lined jaws snap shut. Small prey are killed instantly and swallowed whole, but large prey are grabbed and pulled into the water. Many of these prey are too large to swallow. So, after they have been drowned, their carcasses must be dismembered into bite-sized portions. To accomplish this, the crocodile grabs a chunk of its victim and then spins rapidly on its axis, twisting off large pieces of flesh. The conical, pointed teeth of crocodilians are not suitable for chewing or shredding; they simply swallow the whole mouthful they have twisted off.

Because crocodilians' tongues are fused to their lower jaws they are capable of very little movement, swallowing by lifting the tongue is impossible. Crocodilians take their mouthful, raise their heads to the surface, and then literally throw the food down their throats with a series of jerky head movements. After swallowing, they take a few deep breaths and then submerge for another mouthful.

Although most attacks are launched at the water's edge, crocodiles, despite their short legs, can move on land at considerable speed, albeit only for short distances. Sometimes on land they use their muscular tails or their bony, rock-hard heads as flails to knock their prey down. Such blows are strong enough to knock down most large mammals. The usual strategy is to drive the prey into the water, where the reptile enjoys a significant tactical advantage.

It is thought that hunting by crocodilians is primarily a visual process. Their eyes have large numbers of rods in the retina, allowing them to distinguish black-and-white shapes of prey more than 9 meters (30 feet) away in dim light and to detect quick movement at distances of 30 meters (100 feet) or more.

Jaws 4—Marine Mammals

Millions of years ago, after the great extinction of the Cretaceous-Tertiary boundary, two groups of terrestrial, warm-blooded, air-breathing mammals adapted to a marine existence. These were the pinnipeds, namely walruses, sea lions, and seals and the cetaceans, which include the dolphins, porpoises, and whales. Their modern-day progeny range the oceans, from the freezing waters of the northern and southern polar ice sheets to the equator. All are efficient hunters. In the process of adapting to the specialized requirements dictated by their environment, they became completely divorced from their previous terrestrial existence. They reassumed the sleek torpedo shape of other aquatic

Sonar in the Seas   A number of marine mammals, such as this bottle-nosed dolphin, generate sounds and are able to detect echoes reflected from prey in order to home in on them with amazing accuracy. They also generate a variety of sounds to communicate with other members of their species and can use echolocation to navigate in obstacle-strewn environments.

creatures, retained air breathing, and developed a thick covering of fat, or blubber, to insulate themselves.

The Pinnipeds

The pinnipeds seem to have evolved from some early doglike carnivore. In their return to the sea, they evolved along their precursor's lines. They became the eared seals often referred to as sea lions, the earless seals ranging in size from a few feet to 20-foot-long elephant seals, and the walruses. Intelligent and easily trained, the eared seals are superb, highly maneuverable swimmers that use all four limbs and spend some time on the beaches. All three types are predators, with the seals' feeding mainly on fish, while the walruses' main diet is shellfish. There are, however, reports of walruses' occasionally killing small whales with their tusks.

The Cetaceans

The cetaceans apparently evolved from a terrestrial horselike creature and subsequently developed carnivore-like teeth. One of the most primitive whales was a 75-foot-long Eocene giant known as zeuglodont. This toothed predator looked less like a whale and more like what we imagine a sea monster might have been.

Blue Whales The cetaceans include the largest animals that ever lived on this planet, the great blue whales, *Balaenopter musculus,* the largest of which are 100 feet long and weigh over 140 tons. Like all marine mammals, these giants underwent drastic modifications in their bodily architecture and physiology in the process of adapting to an aquatic existence. Obviously, such monstrous bodies need a vast supply of nutrients. Usually large predators feed on large prey to meet their energy needs; but the blue whale and its cousins among the mysticeti feed exclusively on small, shrimplike animals called krill. Obviously, such small prey must exist in vast numbers, and feeding on them requires a highly specialized apparatus. Instead of teeth, the mysticeti have bony plates extending downward from the roofs of their cavernous mouths. The edges of these plates form a collection of stiff, fibrous bristles that act as filters.

These giants, mouth agape, surge into a school of krill, filling their oral cavities. The jaws shut, the huge muscular tongue lifts and acts as a piston to force the water through the sievelike array of fibers called baleen, trapping the krill, which are then swallowed. Considering that the whale consumes several tons of food per day to maintain itself, its consumption of krill numbers in the billions. However, there are other cetaceans that are efficient carnivores, preying on larger animals. These are the toothed members of the clan, the odontoceti.

Toothed Whales Toothed whales have long captured the imagination of humans, particularly the sperm whales of *Moby Dick* fame and the sleek, big-brained killer whales. Toothed whales communicate with acoustic signals, whistles and clicks at frequencies within the range of human hearing. Each pod has its own unique call signals, but there is some evidence that different pods can and do communicate with each other.

Sound generation is used not only for communication but also for locating underwater obstacles and prey. To accomplish this, odontocetis generate a sound pulse from a bulbous structure in their forehead. On hitting an object, the sound is reflected back toward the source as a weak echo, and this echo is detected by the odontocetis' acutely sensitive hearing organs. Differences in the intensity and arrival time of the echoes at the two ears allow the hunter to determine the location and distance of its target.

We terrestrial animals generate sound by moving air over our vocal cords causing them to vibrate and displace air into a wave of peaks and troughs. The odontocetis have a totally different system in which they compress their air-filled nasal ducts, forcing air over fleshy liplike structures and causing them to vibrate and create high-energy sound waves. These waves are then transmitted through the oil-filled melon in their forehead. The melon apparently acts as a reflector or a focusing device, converting the sound wave to a precise narrow beam. Such a focusing mechanism permits the hunter to detect even very small targets. The most massive of the oil-filled melons is found in sperm whales, whose heads make up almost one-third of their 50-foot length. The huge oil-filled complex structure, the spermaceti organ, has a powerful muscle layer stretching from its front to the top of the huge skull. The sudden contraction of these muscles generates a surge of pressure in the air ducts producing a very high intensity burst of sound.

Sound is measured on a logarithmic scale whose units are called decibels. One decibel is the lowest limit of human hearing, speaking normally registers about 50 decibels, and the sound of a shotgun blast exceeds 150 decibels. Recent studies have shown that a number of cetaceans produce very loud sounds. While the measurements are not precise, we know dolphins can generate sound pulses of about 225 decibels, and it is estimated that sperm whales can exceed this, producing pulses whose intensity may reach 270 decibels.

What function do these loud sounds have? It has been suggested that such intense sound pulses are used as weapons to stun or disorient fish. We can observe the effects of intense sound in the laboratory. For example, the California pistol shrimp is equipped with outsized pincers that can snap together and create a sound so intense that it can stun and immobilize nearby fish. There are also anecdotal reports of fish in the vicinity of echo-locating dolphins becoming disoriented and lethargic. But the best evidence of high-intensity sound being used as a weapon comes from some compelling observations of sperm whale feeding.

Precision Sound   Many whales, such as this one-ton beluga, have bulbous, oil-filled structures, called melons, at the fronts of their heads. These melons allow them to focus a beam of sound with considerable accuracy.

Sperm Whales Sperm whales are extraordinary hunters that seek prey from near the water's surface to a depth of 1,000 meters or more. They usually cruise along slowly at about 3 miles per hour and can, if need be, accelerate up to 15 miles per hour. The weapon systems of the sperm whale are some impressively long, pointed, spaced teeth in its slender lower jaw. These teeth don't appear for several years. Nevertheless, an analysis of the stomach contents of sperm whales reveals a remarkable diversity of prey, ranging from small, slow-moving fish to large, fast-swimming fish like salmon, barracuda, and tuna. One of their favorite foods seems to be cephalopods, cuttlefish, and squid whose indigestible beaks are often found in their stomach.

Among those remaining cephalopods beaks are those of the deep-dwelling mysterious giant squid. Whalers have frequently found scars of sucker marks on the whales' skin. However, none of these diverse prey showed tooth marks, and they appear to have been swallowed whole. How could sperm whales do this? How could they capture large prey like tuna and giant squid that at top speed can travel more than 30 miles per hour? The only logical answer to these questions is that they must stun or kill their prey with a focused pulse of very high intensity sound.

Killer Whales Of all the cetaceans, dolphins seem to have captured the most human interest. The largest of the dolphins, the killer whale, *Orcinus orca*, occupies the position of favorite top carnivore. These streamlined, highly intelligent social predators, weighing up to 6 tons and measuring almost 30 feet in length, are truly awesome hunters. Like the other oceanic top gun, the white shark, they prefer high-calorie fatty diets such as pinnipeds and other whales, including, on occasion, the gigantic great blue whales. But their hunting techniques are quite different from those practiced by the great white shark. Thanks to the recent work of many researchers and some dramatic videotapes, we are now able to reconstruct a reasonably accurate picture of how they hunt.

A pod, often a family group, usually cruises along in formation, with the tall black dorsal fin of the males clearly visible. Suddenly, as if on command, they spread out in a crescent-shaped line and head toward the shore. The pod appears excited; and some of its members leap from the water as they lock in on their next meal, a herd of sea lions. The whales seem energized by the joy of the hunt. Almost playfully, they prod and bump the sea lions that are now fleeing for the safety of the shore. But there is a method to this game. The sea lions, now beginning to tire, are being rounded up into a tightly packed group, as they swim for the safety of the shore. Abruptly the game ends and the killer whales begin to slaughter the herd. (A single bite of a whales' long, intermeshing conical teeth is sufficient to disembowel or kill a seal.) As soon as one kill is made, the whale goes after another and repeats the process, and then again and again. Once the killing orgy is over, the pod begins to feed leisurely and, once sated, reverts to its cruise formation.

Clearly such coordinated hunts require communication and training. Like many land predators, young killer whales go through a long apprenticeship, usually under the tutelage of their mothers. They have learned some truly amazing tactics. In the Antarctic, for example, the whales frequently rise out of the water, using their eyes to scan ice floes for resting seals. This behavior is

The Emperor of Dirty Tricks   While on an ice floe, penguins are almost totally safe
from aquatic predators, but they have to enter the water to feed. Often, as they stand at
the edge of the ice floe, one will nudge a companion into the water. If this unfortunate
fellow is not attacked, the rest of the penguins will dive in.

called spy hopping. If a seal is spotted on an ice floe, the whales dive in formation, burst out of the water near the floe, and create a wave that tips the floe so
that the seal slides into the sea, where it is quickly dispatched. Sometimes they
use the same tactic to push larger penguins into the water, where they are swallowed whole. Penguins seem alert to the presence of killer whales; they usually
won't go into the water when they see spy-hopping whales or tall black fins.
But they never know what's lurking beneath the surface. They gather at the
edge of the floe, seem to ponder, and then dive in en masse, possibly thinking
there is safety in numbers. Penguins also play a cruel survival game in which
one unwitting player is bumped off the ice floe to literally test the waters for
the presence of predators: If the victim swims to the surface, it means all is
clear and the rest of the penguins dive in.
    Killer whales' most remarkable hunting technique involves a brief foray
onto land where they deliberately beach themselves to capture prey at the
shore's edge. The orca targets a prey, usually a pinniped, makes a high-speed

dash toward the shore, beaches itself, snatches the seal in its mouth, and then rocks itself sideways toward the incoming waves that wash it and its prey back into the sea. Sometimes this beach hunting is done simultaneously by several members of the pod in a cooperative effort. Often the orcas play with their wounded prey, throwing it high in the air, sometimes swatting it with their tails until they tire of the game and finish their meal. What the function of such play is remains a mystery.

Killer whales are very catholic in their tastes. They gulp down salmon and penguins whole, they devour seals and sea lions, and there are even reports of them plucking moose off the shoreline. Most astonishing is the fact that they attack great blue whales that weigh over a hundred tons. Of course, one killer whale can't do this alone; but large pods can. They surround the big blue on the flanks, fore and aft; some position themselves below to stop the whale from diving to escape; they may even try to prevent the blue from surfacing to breathe while other members of the pod shred the blue whale's tail fins to limit its swimming capacity. This array of tactics attests to the killers' ability to learn and remember. During the coordinated attack, which continues for hours, the great blue is literally eaten alive, as members of the pod come in relays to rip great pieces of blubber off its sides. Once the whale is dead, the feeding continues at a leisurely pace without competition from fellow pod members.

Orcas are also known to hunt and kill other toothed whales. In 1998, a U.S. National Oceanographic and Atmospheric Administrator (N.O.A.A.) research vessel, off the California coast, watched as a pod of nine, 30-foot-long female sperm whales were subjected to a relentless and devastating attack by killer whales. The defensive strategy of the sperm whales was to form a circlelike rosette in which their heads were pointed inwards and their powerful tails were pointed outward. A blow from one of these huge, muscular tails is capable of seriously wounding an attacker. Although smaller and lighter than the sperm whales, several orcas, all females, after circling the rosette, charged in, broadsiding the sperm whales and inflicting terrible wounds. The orcas would then withdraw for several minutes, regroup, and charge again. Almost shoulder to shoulder, four adult female orcas in a highly coordinated charge leaped high out of the water just before they slashed into the pod. Each attack tore mattress-sized slabs of blubber from the sperm whales' flanks. After a while, two killer whale calves accompanied the females orchestrated charge suggesting that this was, for the youngsters, a training exercise. These attacks continued for three hours and the sperm whale defensive formation began to break down.

One severely wounded sperm whale drifted off, and two of her sisters came to her aid, flanking her and trying to lead her back to the circle. But, the would-be rescuers were assaulted and grievously wounded. The defensive rosette broke down, and more and more orcas were recruited to the area. The big male orcas with their distinctive, tall, dorsal fins had been waiting nearby but had not attacked. Finally, a large bull joined the battle to administer the coup de grace and drag the sperm whales' carcasses away to be skinned by dozens of other orcas that had answered some unheard mess call.

Clearly, the hunters' orchestrated team tactics were learned through long practice and were honed to perfection by repeated earlier experiences. Such behavior would be expected in such a big-brained, social animal. On the other hand, the equally big-brained sperm whale's defensive strategy was totally ineffectual. Sperm whales can dive deeper and stay submerged longer than the orcas, and their toothed lower jaws are capable of inflicting killer wounds. Why they did not learn to disperse and dive is inexplicable.

What is amazing is that these voracious predators have never been known to attack a human. They and other dolphins even seem to enjoy the presence of humans. Killer whales in captivity are very receptive to training and even like to show off. But who knows what these great dolphins think? Sometimes I question the morality of imprisoning such intelligent creatures for entertainment and profit. The U.S. Navy certainly abused dolphins, training them as dispensable underwater submarine hunters. At one time, the navy attempted to use the dolphin's extraordinary hearing and intelligence to recognize the sound of enemy submarines, home in on their target, and then strike it with an attached explosive charge, destroying the sub, and the dolphin as well. Fortunately, this method was never used, but the navy did train dolphins to locate expended torpedoes so that they could be recycled. For years, researches have tried to learn the language of dolphins in order to probe their thought processes; but to date, the voice of the dolphin remains an undecipherable mystery.

Missile Systems of Aquatic Predators

In the brackish waters of river estuaries in Southern Asia, Australia, and the East Indies, live several small fish of the genus *Toxotes,* who use just such a tactic. These fish have developed a most peculiar weapon system, a sort of submarine-launched ballistic missile. More correctly, this system should be called a subsurface-to-air missile system since the targets are flying insects and the missiles are water droplets launched from the fish's mouth. Because of the surface tension properties of water, it forms droplets when in the air. These droplets have enough mass to stun and bring down an insect. Once the victim has fallen into the water, the archerfish closes in and gulps it down. Droplets are fired in salvos, which increase the probability of multiple hits. The fish's optical system, which does not project from above the surface, must compensate for the refraction (bending) of light as it passes from air to water. Furthermore, its "onboard computer," its tiny brain, must be able to calculate the trajectory of the set of droplets. Under good conditions, Archerfish can hit targets at altitudes of about 20 inches.

Torpedoes and Missiles of Microscopic Aquatic Predators

In the aquatic battlefields are vast armadas of primitive, microscopic hunters. Some of these miniature dreadnoughts are heavily armored, with ornate shells made of calcium, silicon, cellulose, or protein. Some of the armored, single-celled, flagellated hunters are equipped with potent stinging cells, nematocysts similar to those of the Portuguese man-o-war and propel themselves with one or more long whiplike oars, or flagella. Some of these armored flagellates also produce a highly potent nerve poison that sometimes becomes concentrated in clams, causing paralytic shellfish poisoning, or in game fish, causing ciguatera (poisoning) in anyone who eats them. A number of other single-celled protozoa are highly efficient predators capable of pursuing, immobilizing, and devouring prey as large as themselves. Among the most voracious of these carnivores is *Didinium nauutum,* a ciliated hunter that feeds exclusively on paramecia.

Both *Didinium* and paramecia propel themselves through the water by means of short, contractile hairlike structures called cilia. Each cilium is an engineering marvel. In paramecia, the cilia are arranged in longitudinal rows, and the beat of the cilia is highly synchronized. Viewed under a microscope, they whiz through the visual field at apparently enormous speeds, and indeed, in their microcosm they are exceptionally quick; but in our world it would be less than 1 yard per hour. Of course everything is relative: a paramecium is only about 100 microns (1/250 inch) long.

Between the cilia and imbedded in their outer coat are batteries of miniature cannon-like structures called trichocysts which apparently can be discharged on command. When fired, each trichocyst releases a long, banded thread tipped with an arrowhead-shaped sticky barb. These harpoon-like munitions not only immobilize the prey but also serve as a means of attachment. The paramecium killer, *Didinium,* is equipped with even heavier armament. In its proboscis, are batteries of giant undischarged trichocysts, which are toxic. When *Didinium* makes contact with a paramecium, it repeatedly jabs with its proboscis at its target and discharges its trichocysts, which rapidly immobilize the doomed paramecium by paralyzing its cilia. *Didinium* then opens its proboscis to an enormous gape and swallows its victim whole. It

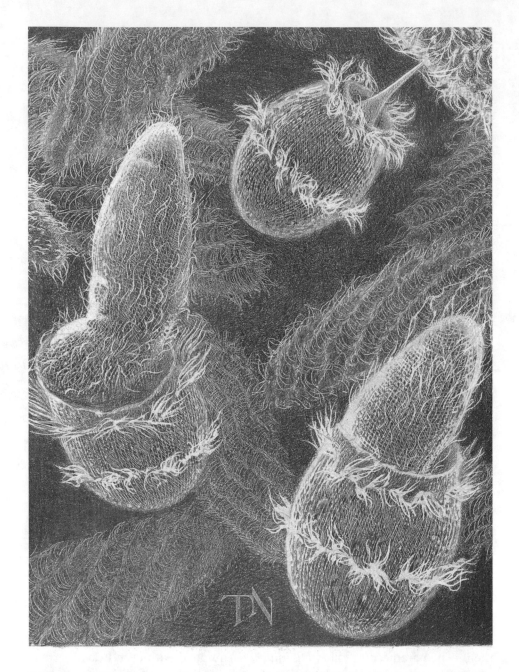

Underwater Missile Systems The ciliated hunter *Didinium* preys on paramecia. When attacked, a paramecium will launch a barrage of sticky missiles called trichocysts, but these don't deter *Didinium*, whose own giant trichocysts are toxic missiles that paralyze the paramecium, which is then swallowed whole.

hardly seems imaginable that such modern day life-and-death encounters as these involve weapon systems that evolved well over a billion years ago.

The missile system used by the ciliated protozoan is quite primitive compared to the weaponry of the microscopic parasitic fungus *Haptoglossa mirabilis,* which preys on wheel animalcules, or rotifers. Rotifers are microscopic multicelled animals whose round, ciliated crowns give them the appearance of having revolving wheels. Some rotifers are mobile; others are sessile. Most are omnivorous and create feeding currents with the cilia; others actively detect, pursue, and capture prey with a protusible, forceps-like weapon. However, rotifers are also preyed on by many organisms including *H. mirabilis* with its extraordinary missile system. These so-called gun cells are shaped like short-barreled miniature cannons. The missile chambers of the cells are anchored to an adhesive base, allowing the barrel of the gun cell to be raised at an angle that increases the odds of an encounter with a passing rotifer. When contact with an appropriate target is made, the gun cell fires its harpoon-shaped missile, an encysted spore which penetrates the rotifer's armored pellicle. Once inside the rotifer, the injected spore immediately initiates an infection by the parasitic fungus, which at once starts to make new fungi.

Missile Systems of Passive Hunters

One of the earliest and most successful hunting strategies was one in which the predator did not have to expend energy in stalking and pursuing a prey. Some of these predators were sessile, attached to rocks or some other substratum on the bottom, and others were floating, their movement dependent on winds and ocean currents. What all of these organisms had in common was an extraordinary weapon system composed of vast batteries of missile-firing cells. These missiles were loaded with highly toxic, quick-acting poisons, some of which are among the most deadly compounds known to man. This weapon system evolved early (700 million years ago) in the history of multicellular organisms in a phylum known as Cnidaria (Gk "nettle").

Found in both fresh- and saltwater, cnidaria range in size from the microscopic to giant jellyfish whose bell can be 10 feet in diameter and whose sting-laden tentacles are 100 feet long. Some float on the surface and have saillike structures, some float below the surface, some are found in the deepest parts of the ocean's abyss, and some are attached to fixed objects and never move. They can be divided into three groups: the anemones, the jellyfish, and the corals. All share one significant characteristic: they have specialized cells, called nematocysts, that can fire out thin threads to either snare or sting their prey.

The cnidaria developed a veritable arsenal of diverse types of nematocysts, some with short tubes, some with long tubes, some with smooth tubes, and

The Deadly Sea Wasp   The missiles of the sea wasp, microscopic toxic darts called nematocysts, contain one of the most potent poisons known.

some with barbed tubes. Each nematocyst consists of a capsule within which a hollow thread is coiled—the loaded munition. Some are equipped with a pressure-sensitive projection, a sort of trigger mechanism called a cnidocil; and others are triggered by contact with a specific chemical on the prey's surface. The toxic nematocysts have a poison-filled sphere attached to the tube. The venoms in this sac are a mix of toxic proteins some of which block ion channels in muscle and nerve cells. A single species may have as many as ten different toxins. Given the fragile nature of the jellyfish and its vulnerability to damage, it would never do to have a prey thrashing around in protracted death throes. Therefore, the evolution of a highly potent fast-acting toxin was essential.

The deadliest of the jellyfish is the box-shaped *Chironex flerkeri,* which is without question the most venomous sea creature known and is commonly called the sea wasp. A large sea wasp has enough venom to kill sixty adult humans. Off the beaches of Australia, sea wasps account for more human deaths than do the great white sharks that abound in the same waters. A fully grown sea wasp has a square-shaped translucent bell that pulsates and, in so doing, can propel the animal at speeds of up to 5 feet in 1 second. However, such bursts of swimming are rare and are not used in pursuit of prey but in evasion of predators. Each quadrant of the sea wasp has a photoreceptor complex complete with a lens, but little is known about how this works. The basketball-sized body contains a gut with a well-defined mouth and reproductive organs. Along the margins of the bell hang some sixty tentacles up to 15 feet long and loaded with thousands of cocked, stinging cells. It is obvious that a double trigger mechanism must be involved. The pressure trigger, or cnidocil, must first be cocked by the presence of a chemical signal from the prey's surface; otherwise every accidental contact with an inanimate object such as a rock would trigger the salvo. Once triggered, the barb-tipped thread is fired in less than 3 milliseconds with an enormous acceleration velocity, apparently propelled by hydrostatic pressure.

If a *Chironex* missile salvo accidentally strikes a large animal such as a human, death can occur in less than 4 minutes, since one of the toxins blocks calcium channels in heart muscle. Other toxins are carried to the brain and block the respiratory centers, and yet a third toxin bursts the membranes of oxygen-carrying red blood cells. Victims describe the pain caused by such stings as excruciating. This animal, whose substance is like a Baggie filled with Jell-O, kills faster than the most venomous snake. However, even this most deadly of marine organisms is also prey, to the hawk-billed marine turtles that are apparently immune to the box jellyfish's deadly mixture of venoms.

The most commonly encountered stinging jellyfish is the conspicuous Portuguese man-of-war, or blue bottle, found floating by the tens of thousands off the southern Pacific, Atlantic, and Gulf coasts. The Portuguese man-of-war,

Physalia physalis, is not a single jellyfish but a colony of individuals hanging from a buoyant translucent bladder tinted with blue, pink, and purple and filled with nitrogen, oxygen, carbon monoxide, and other gases. The float is from 5 to 15 inches long and may project out of the water 7 to 8 inches. (*Physalia,* unlike the single jellyfish, can't swim, but is completely at the mercy of the currents and winds.) A tangle of long tentacles hangs below the float, is bearing colonies of reproductive individuals, digesting individuals, and sting-ing individuals. Stinging cells, large and small, are grouped in batteries arranged like rungs on a ladder, along the entire length of the tentacle, and each tentacle may contain three-quarters of a million nematocysts.

A small portion of the *Physalia* nematocyst's membrane is folded back into its interior to form a long, coiled, hollow tube equipped with sharp barbs or hooks at its base. The entire nematocyst is nestled within a second fluid-filled capsule called a cnidoblast. The cnidoblast has a cover which acts as a door that is spring-loaded shut. When touched, pressure-sensitive, trig-ger-like spine projections from the cnidoblast cause the door to open. The coiled thread tube of the nematocyst is forcibly discharged, its sharp hypo-dermic-needle-like end punctures the skin, and its hooks and barbs lock it in place. Then the venom-containing sac contracts and injects the poison into the skin.

A single brush with a tentacle may produce thousands of stings. A salvo of nematocysts is often lethal to any fish that accidentally blunders into a tentacle. But there are some fish that are apparently immune to the toxins and swim within the tangle of tentacles. The reason for this behavior is not clear, but it has been suggested that this is a defensive strategy that affords the fish immune to the toxin protection from other predators. In fact, the use of stinging cells as a borrowed weapon system is surprisingly common. The colorful, sessile sea anemones don't swim or float but instead spend their lives latched onto some surface. They make their living by stinging prey that blunder into their crown of tentacles. Often found in this maze of deadly tentacles are colorful clown fish. Immune to the stinging cells, these fish share the scraps of *Physalia*'s meal and may lure other fish into the ten-tacles the anemones use to kill.

A species of crab, *Melia tesselata,* that inhabits the coral reefs, recruits the stinging powers of cnidarians by grasping the soft body of anemones in their pincers and holding them in servitude. The crab can then use the anemone tentacles to deter predators or to capture prey. This symbiotic relationship obviously benefits the crab, and at the same time, the crab's motility and prey-location systems enhance the anemone's chances of getting fed.

Defense and weapon systems are borrowed. Some species of hermit crabs use the shells of deceased snails as armor protection against predators and also cover their borrowed armor with anemones. These crabs pluck anemones from

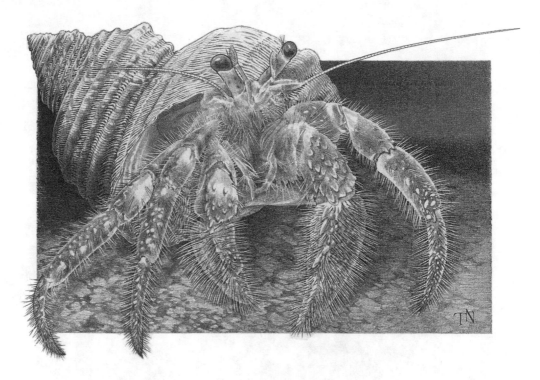

Living on Borrowed Defense The soft-bodied hermit crab acquires protective armor without having to expend the energy needed to produce a shell. It simply finds the shell of a deceased snail appropriate to its size and anchors itself within. If attacked by a predator, the crab contracts its abdominal muscles and withdraws into the shell. Some hermit crabs augment their defenses by anchoring onto their shells sea anemones whose tentacles are loaded with potent stinging cells that deter predators. This is a symbiotic relationship since the nonmotile anemones benefit by being constantly moved to new feeding grounds as well as from the food fragments of the crab's meals.

the perches with their foreclaws and then hold them against the shell until the anemones attach.

The most elaborate use of borrowed stinging cells is seen in the naked marine snails called nudibranchs. Without the protection of a shell, these slow soft-bodied animals would seem to be highly vulnerable to predation, but they have acquired a potent defensive system which they advertise. Some species of nudibranchs are brightly colored and quite handsome. Their beauty warns predators, "Don't eat me; I'm toxic." These toxins are acquired by eating cnidarians and somehow appropriating their prey's stinging cells, storing them in finger-like processes that

Anemones   Anemones such as *Metridium senile* are cylindrical, immobile hunters that attach themselves to solid objects. Around their oral discs are contractible tentacles, loaded with stinging cells which immobilize prey that blunder into them. These tentacles contract to carry the immobilized prey into the anemone's gullet. Floating in the background are large disc-like jellyfish, *Cyanea,* whose long tentacles are also heavily armed with stinging cells. Some large *Cyanea* have tentacles over 30 feet long.

cover the back of the snail. How the nudibranchs manage this is not understood, but the end result is a highly potent defensive armament for the snail.

Advanced Missile Systems

Missiles systems give predators the ability to hit their prey from a distance. This is a significant advantage for the predator because many prey species have a programmed critical distance. On entering the prey's critical space, any predator elicits an escape response; so being able to hit a prey from a distance has obvious advantages.

The beautiful cuttlefish and some of its cephalopod cousins have an optically aimed missile system that gives them the capacity to hunt at long range. Two of their tentacles, the raptorial tentacles, are fired like ballistic missiles. The firing of the tentacles is produced in part by elastic recoil of a compressed rubber-like protein and in part by the contraction of fast-twitch muscle fibers.

Another long-range weapon system used by aquatic hunters is their rapidly protrusible tooth-lined jaws. The adult sling-jaw wrasse, *Epibulus insidiator,* of the Indo-Pacific coral reefs, is a colorful, medium-sized predator that can rapidly protrude both its upper and lower jaws more than one-third the length of its head.

One of the most efficient optically aimed missile systems is found in amphibians, such as frogs and one very interesting species of salamander. While these animals spend part of their lives on land, they also spend a lot of time in the water; for that reason, they are included in this chapter.

Many species of frogs have a ballistic tongue that is projected out of their mouth by muscular action. The tongue is optically aimed, a sticky pad on the tongue's tip captures the insect prey, and the tongue with its prey is reeled in

What the Frog's Eyes Tell the Frog's Brain   The leopard frog's eyes can detect both form and motion. But such information is only partially processed in its eyes; signals are subsequently sent to its brain, where they are processed to guide the frog's optically aimed tongue toward its prey, much like a guided missile.

quickly. (Later in the book, we will discuss a similar system used by terrestrial lizards, or chameleons.)

The ultimate ballistic tongue is found in salamanders of the genus *Hydromantes*. Unlike the tongues of frogs and chameleons, it is partly skeleton and extends into the body cavity as far as the pelvis. When not in use, the bony missile system is folded neatly in the salamander's mouth. When a potential meal is spotted, the mouth opens like a silo door and the tongue is launched like a missile by muscular contraction. The sticky-tipped projectile reaches its target in a few milliseconds, and then the tongue and insect are recoiled into the mouth by powerful muscles located near the salamander's rear limbs. The range of this missile system which remains attached to the salamander's body is exceptional. It extends to 80 percent of the animal's length.

Missiles from the Seafloor

When one thinks of dangerous predators, the last thing to come to mind is a slow-moving snail. Yet some 350 to 400 species of the lowly marine mollusks, the hunting cones, are truly prodigious predators, capable of killing fish that swim much faster than their own snail-paced locomotion. These nocturnal fish eaters of the genus *Conus* are hunters of the seafloor and of coral reefs, where they most probably detect their prey's chemical essence. The key to their success is a formidable missile system that fires venom-laden darts whose toxins (conotoxins) hit the victim's nervous system, immobilizing it within seconds. The missile launcher, the snail's serpentine proboscis, contains a dozen or so needle-shaped, hollow barbed darts called radulae. These darts are propelled out of the proboscis by a hydrostatic pressure created when the snail contracts a muscular bulb; only in this case, the hydraulic fluid propellant is the highly toxic venom. Spare radulae can be loaded onto the launch platform in about 10 minutes or so. There are a number of different conotoxins, some much more potent than the most poisonous snake venom, and there are several confirmed cases of human deaths caused by cones. In the usual scenario, a shell collector picks up one of these handsome shells and is rewarded with a trip to the mortuary.

Chemical Warfare at Sea

The term *poisonous* refers to organisms that contain toxic substances in their tissues and are otherwise harmless unless eaten. Obviously, this is not a good survival strategy for the individual that got eaten, but it is possibly good for its kin. Some animals are poisonous by accident, because they have eaten something which is toxic to others but to which they have immunity. Two of these accidentally toxic groups of organisms are shellfish, such as clams and mussels, and bony fishes, such as barracuda and jack. Both acquire their toxins from flagellate protozoans, algae that periodically undergo huge population booms when nutrients become available. Being filter feeders, clams suck in and eat large numbers of red-tide organisms, or dinoflagellates, when red tides containing billions of these microscopic toxic algae are washed ashore. When this happens, the otherwise edible shellfish become deadly to eat. The toxic agent, saxitoxin, is a nerve poison that blocks ion channels in nerve and muscle cells and causes respiratory paralysis. Clams and mussels are probably responsible for more human deaths per year than are highly touted killers such as sharks.

The other common fish poisoning is called ciguatera. The chemical nature of ciguatoxin is unclear, but it seems to block an enzyme that gets rid of the neurotransmitter acetylcholine. So its effect is very much like that of the nerve gas used

Chemical Warfare at Sea Many marine creatures have evolved potent chemical defenses and toxin-delivery systems. Tiny, armored, single-celled, red-tide organisms possess toxins which kill millions of fish, and indirectly, some humans who eat larger organisms that have become poisoned. The bite of the small blue-ringed octopus can be lethal to humans, and the puffer fish's toxin, contained in their skin, liver, and ovaries, is responsible for a few dozen human deaths per year.

by Iraq's Saddam Hussein against the Kurds. The source of this toxin is a tiny dinoflagellate, *Gambierdiscus toxicus,* found in local coral ecosystems in small numbers. Sometimes after a storm, the waters contain lots of nutrients for these dinoflaellates and the population explodes. The small reef fish feeding on them become toxic. The little fish are eaten by larger fish, like barracuda, groupers, snappers, and many other favorite edible game fish, which accumulate the toxin. These big fish become very dangerous to eat. Indeed, they have caused many deaths.

The most notorious of poisonous fishes are the puffer fishes and their kin, which are responsible for hundreds of deaths every year, particularly in Japan, where they are considered a delicacy. The poison contained in these fish, tetradotoxin, is one of the most dangerous fast-acting toxins known; the mortality rate in humans ingesting it is about 60 percent. Tetrodotoxin, or TTX, has been extensively studied. Its main action is to block sodium ion channels in nerve cells and render them inoperative.

A number of other advanced marine organisms have evolved toxic chemical weapons of their own and are actively venomous. They possess a venom apparatus consisting of a delivery system, or stinger, and its associated venom-secreting glands. Such a weapon system can be used to subdue prey or as a defense to discourage predators.

Among the earliest actively venomous fish were the stingrays, which belong to the class Chondricthyes (meaning fish with skeletons of cartilage). Stingrays come in all sizes, the largest being 3 feet in diameter and 12 feet long (with most of the length in the tail). They live in shallow waters, usually in saltwater, and spend most of their time on the bottom. Masters of camouflage, they often burrow into the sand bottom where their flattened bodies blend with their surroundings. Their weaponry is purely defensive consisting of an erectile stinger located on the upper side of the tail. The bone-hard stinger is a flattened, sharply pointed weapon, whose sides are lined with sawtooth-like barbs. Because such structures have significant capacity to cause serious wounds, a number of native tribes of the Caribbean have utilized stingers from dead rays as spear points. In the living ray, the stinger is covered with a sheet of skin that contains venom glands. When disturbed, the ray raises the stinger (some rays have more than one) and, with a flick of its long, slender muscular tail, drives the stinger deep into the source of its annoyance. When the barb is withdrawn, some of the skin and glands are left in the wound. The venom is a mix of several toxins and causes excruciating pain and tissue damage. Very rarely, however, is it lethal in man. While the rays actively deliver their sting, most venomous fishes are passive stingers requiring a predator to make contact with venomous spines attached to their fins or gill covers.

There are several groups of venomous fish, the most notorious and dangerous of which belong to the family Scorpionidae (scorpion fish). Some living in the shallow waters of coral reefs are among the most bizarrely shaped and spectacularly colored fish known. Their needle-like spines, located on the dorsal side, pelvic area, and anal area, are covered with epidermal sheaths containing the venom glands. Envenomation is the result of physical contact, and the fast-acting poison causes immediate and intense pain, making it an effective deterrent to predators.

There are a number of toxic scorpion fish, but one group deserves special mention because their toxin is often lethal. They are *Scorpaernidae horrida* and *Scorpaernidae trachinus,* both referred to as the deadly stonefish. These sluggish, grotesquely ugly bottom dwellers have bizarre forms and colors so that they blend in with the bottom. They dig themselves in and wait, undetectable, to ambush their prey. Their stinging equipment is purely defensive; it consists of sharp, short, thick spines containing venom glands that produce a lethal toxin. Untreated envenomation of humans is fatal about 70 percent of the time.

The only other marine animals that actively envenomate are snakes of the family Hydrophiidae all of which are venomous and all of which are totally

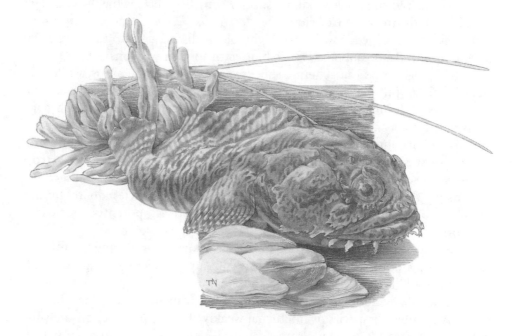

Don't Tread on Me!   Several bottom-dwelling bony fish, such as the toadfish shown here, are equipped with pointed, hollow spines attached to poison glands. The poisons contained therein are among the most potent nerve toxins known, and stepping on the deadly toadfish can be a terminal event.

adapted to marine life. These sea snakes are found, sometimes in huge numbers, off tropical coasts in the Pacific Ocean. They use their venom to immobilize their prey. Not having hypodermic-like fangs, as do many other venomous snakes, they must bite their venom in. During such a bite, the snake usually releases all of its venom to assure quick knockdown of the prey. They are usually not aggressive to humans. However, they will bite defensively if provoked. All sea snakes contain enough venom to kill many humans. In the few humans that have been bitten, the mortality is very high.

Electric Weapon Systems

Water not only conducts sound well; it also conducts electric current. This property has been exploited by a number of fish with some ingenious adaptations that allow them to generate electric fields to help them find prey; detect predators; navigate in murky, obstacle-filled waters; and even communicate with each other. Seven families of fish have evolved systems that generate electric fields strong enough to stun or even kill prey. Both bony fish, such as the 9-foot-long, sluggish South American electric eel *Electrophorus electricus,* and cartilagenous fish, such as the torpedo ray, release high-voltage discharges into the water. No terrestrial or aerial organism uses such bioelectric tools or weapons.

The torpedo rays can generate shocks of 60 volts and 50 amps, while the electric eel can put out stunning pulses of over 500 volts. Their electric organs are stacked columns of structures that are modified muscle cells; each column may have as many as 10,000 of these electroplaques. In the electric ray, electroplaque columns make up about 40 percent of the fish's weight and are located in the ray's wings at right angles to the spine. The arrays of electroplaques are set up in a series, with each plaque's adding bit by bit to the voltage, using the same mechanism as nerve cells and muscles do to pump metallic ions and thereby building up an electrical potential across a membrane.

This accumulated charge can be fired on command, producing a powerful electric field around the fish from head to tail. Any intruder entering this field is stunned or killed.

Many fish, including rays, sharks, and eels, can produce weak electric fields. A number of species of nocturnal weakly electric fish dwell in the silt-laden, murky waters of South American and African rivers—an alien world in which an electric sensor aids survival. The handsome, 20-inch-long *Gymnarchus niloticos* is one. Unlike other fish, *G. niloticos* does not swim by undulating its long, thin, naked tail but by undulating its long dorsal fin. It can execute movements with great precision and avoid underwater obstacles even though it has very poor vision. Apparently, these fish have evolved a

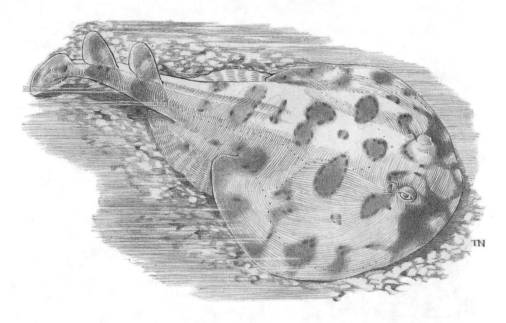

Shocking   The lesser electric ray shown here and several other marine and freshwater fish can generate high-voltage electric pulses to stun and immobilize prey, or to ward off possible predators. The high-voltage pulses of the Amazon electric eel are intense enough to kill an adult human.

novel and remarkable capacity to generate weak electric fields around themselves and then detect any object in that field. Their pulse generators, like those of their strongly electric relatives, are modified muscle cells which discharge up to 300 times per second. (Not all electrolocating fish use muscle-like cells to generate a field; instead, some, like the transparent knifefish, *Eigenmannia virescens,* use nerve cells.)

The tip of the motionless tail acts as a negative pole, the head as a positive pole, creating a symmetrical electric field around the fish. Anything disturbing this field, even a fraction of a millionth of a volt, can be detected. Their supersensitive electroreceptors consist of jelly-filled pores topped by sensory cells and are located in their head. Stimuli from the electrosensors are relayed to their brain, where a large, complex neural circuitry converts them into a virtual electric field search image.

The more we study these remarkable creatures the more awesome their electric discharge system seems. Some show species-specific discharge rates, some apparently use electric discharge in establishing territories and in seeking mates. They even seem to show a level of civility missing in today's human society; for example, one Gymnid will signal and ask nearby fish to

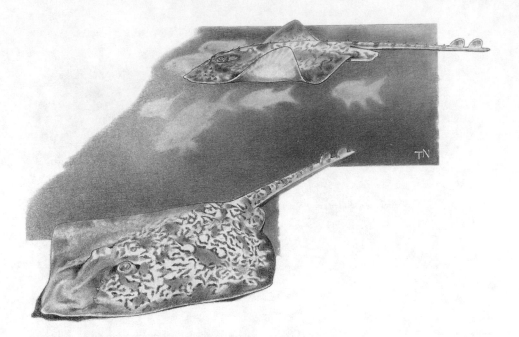

Electrolocation and Communication   Several groups of fish, usually inhabitants of murky waters, generate electric fields around themselves and can detect any disturbance caused by an object within those fields. This is used to locate prey and avoid obstacles. These weakly electric fish, such as the skates shown here (close relatives of sharks and rays), can also communicate with each other by creating volleys of electrical pulses.

stop signaling and they will comply temporarily out of courtesy. Why these fish developed this location and communication system is inexplicable; there are many other species of fish that share their environment and get along quite well without electrolocation. But its extent and sophistication is astonishing.

Defensive Tactics and Strategies

As in other environments, survival in the sea has one prime requisite: speed of operation. In order to successfully detect and evade hunters, prey must have early-warning alarm systems as well as relatively simple, rapid responses to keep them out of harm's way. The alarm system is always a neural process, and the sensory information must be decoded into a startle response that bypasses central processing and analysis. By keeping the startle response to a purely sensory motor circuit, the prey gains speed of reaction.

To be successful, the predator's strike must be faster than the prey's startle time and escape response. Differences of only a few thousandths of a second make the difference between success and failure. This is the oldest game in the world, and in the marine environment there are several fascinating examples of how prey evolved mechanisms to buy time and achieve an advantage.

One widely practiced strategy is to confuse the predator's processing of sensory inputs. Since most underwater predators rely on vision as their primary method of hunting, prey evolved a variety of tactics to nullify the predators' target acquisition, including the evolution of a distinctive color pattern that short-circuits the predators' aggressive behavior.

Schooling Safety   Schooling, or swimming together in large, rapidly moving groups, has proven to be an effective strategy in decreasing the hunting efficiency of visually oriented predators.

Schooling has proven to be an effective strategy that is also based on confusing the predators' visual perception. Fish schools are temporary groups of fish of a given species that are highly organized in formations containing up to a million individuals. The probability of early detection of an approaching attacker is also greater when many individuals are involved. Individuals within the school align with military precision and move in unison, maintaining a fixed spacing and velocity, mainly by visual contact, although different species of fish use a variety of communication signals to coordinate movements.

The school executes elaborate maneuvers and takes on a variety of shapes. It may use one formation while resting, another while traveling, another for feeding, and still other formations to evade attack. To a potential predator, it may appear to be a large organism. When attacked, some schools scatter, presenting the predator with a confusing array of motion; other schools close formation and their members wheel in unison. Still others surround and confuse the predators.

Some fish discourage visual hunters by rapidly changing their size when approached by a predator. Predators are highly selective in the size of the prey that they eat. If their target suddenly assumes dimensions that do not meet their feeding specifications, it will be avoided. A number of puffer fish, aptly named, have the capacity of sucking in enormous amounts of water and ballooning themselves to three times their normal size. Puffing up brings a secondary defensive system into play. Because the skin is drawn taut, their flattened sharply-pointed spines become erect; this is why puffers are also called burr fish. Finally, puffers, particularly during their mating season, produce large amounts of the potent nerve poison TTX, discussed earlier in "Chemical Weapons."

Camouflage is also widely used by prey to avoid detection by predators. Some fish use countercolor shading, being dark and mottled on the dorsal sides and white on the undersides. Look-down hunters can't see them because from above they appear to blend with the darkness of the depths. Look-up hunters, relying on a bright, whitish background of the surface, have difficulty detecting fish with white underbellies. Some fish have luminescent organs which they can control. By illuminating their undersides, they create countercolor camouflage with light.

Many fish evade predators by partially submerging themselves in muddy or sandy bottoms. Some of these fish can also rapidly change colors and patterns so that they blend with their surroundings. But the most bizarre evolutionary visual deception game is practiced by fish that alter their shape to match the shape of a plant. The master of such disguises is the small, 6-inch-long, Saragassum fish whose surface is festooned with skin filaments and tassels that mimic the shape of Saragassum weed, a floating plant. Because of these features, the Saragassum fish is invisible to larger predators. It is also invisible to prey fish and is a voracious predator in its own right. Its large mouth and deep distensible abdomen allow it to swallow fish as large as itself.

Confusing Look-Up Hunters Some deep-sea fish illuminate their undersides to create countercolor camouflage.

The last of the antipredator visual strategies is the evolution of false eye-spots, a strategy not only practiced by fish but also common in a number of other phyla, particularly the insects. Predators always look for eyes, since this tells them where the vulnerable head of the prey is and allows them to predict direction. Very large eyes also serve as a deterrent to attack, since big eyes usually belong to big animals. The banded butterfly fish, *Chaetodon sedentarius,* has very large, conspicuous eyespots on its rear flanks. With this characteristic, it manages to provide directional misinformation to a potential predator.

Another antipredator strategy interferes with the predators' sense of smell. In amongst the coral reefs lives the small, handsome, blue-and-black-striped cleaning wrasse, *Labroides diminuatus,* which services large predatory fish by picking parasites off their surfaces, even on the insides of their mouths. The large fish even seem to wait in line to be treated. This mutualistic relationship brings food to the wrasse and eliminates a source of irritation for the predator. The predator in turn pays for this service by not eating the wrasses. However, night-feeding predators consider the wrasses food. So, in order to survive, the wrasses evolved another antipredator strategy. They find a crevice and then secrete a mucous bag around themselves. The bag keeps their odor in, allowing them to evade detection by the hunter's olfactory sensing system.

Bundling Up for the Night   The cleaning wrasse shown here rests in a coral niche for the night. To protect themselves, wrasses secrete a mucous capsule around themselves, which enables them to evade predators that use chemical sensing to find prey.

Territorial Defense

Many marine organisms have fixed territories whose boundaries they protect with great vigor. Having exclusive rights to a territory ensures the occupant of a number of available resources (space, food, and so on). However, other individuals of the same and other species may want access to the same resource and will compete for it. Such competitions are contests in which there are winners and losers. To the victor go the spoils; the loser must explore new options.

The mechanisms of contest competition are all based on aggressive behaviors—threat, intimidation, assertiveness, and fighting. A cost-benefit analysis of the payoffs to a strategy of competition indicates that if the victor can ensure the availability of resources by establishing the security of his boundaries, it will have been worth the fight. However, if the cost is too high, that is, if it takes time and energy that could have been better applied to other activities, then fighting is not an adaptive strategy. As it turns out, aggression is a fundamental characteristic of all living things and is adaptive, because in competitive contests the victor imposes the probability of his genes' being passed on to the next generation. The bottom line is that natural selection selects for fitness and being fit—winning—favors aggressive genes.

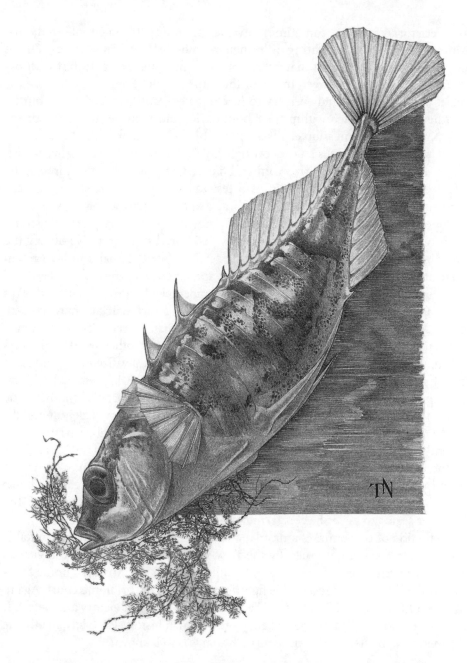

Deceptive Defense   Three-spined sticklebacks are territorial and defend their nest eggs and young with considerable belligerence. They also like to feed on the eggs and young of other sticklebacks. Here, a male, seeing the approach of a group of cannibalistic male sticklebacks, abandons his own nest, moves off some distance, and pretends to be feeding. The intruders, taken in by the guile of the nest protector, will follow, thinking he has found a food source. How such complicated deceptive behaviors have evolved is not known.

Territorial defense sometimes involves fighting off potential predators. More often it involves fighting off a male member of the same species. This is not to say that there aren't many examples of female territoriality, but territorial fights usually are between males of the same species. Frequently, these same males willingly allow and even try to lure females into their territories; but let another male encroach within their boundaries and then, regardless of previous experience, the invasion elicits an attack by the defender of the turf.

The most meticulous studies on fighting behavior and territoriality in fish were done by the Nobel Prize winner, Niko Tinbergen, using the three-spine stickleback, *Gastrosteus aculeatus,* as a test animal. Three-spined sticklebacks are small (4-inch-long) inhabitants of brackish and estuarine coastal waters off Europe, Asia, and Pacific North America. They have three sharp erectile spines in front of their dorsal fins. Males are most territorial and aggressive during the breeding season, when they construct nests. They incite females to lay eggs in their nests, and then they vigorously guard and defend the young.

During this breeding season, the male's underbelly becomes bright red. This red "flag" is used in a number of ways. When any male stickleback literally sees red, it attacks. In a series of studies, Tinbergen showed that the red acts as an "innate releasing factor," a signal that triggers attack behavior. In other words, this fish's neurally based behavior repertoire is hardwired, genetically fixed, to produce a highly predictable, aggressive attack behavior when the fish is presented with an appropriate environmental signal, in this case a red color flash. So, if another male enters the home territory of a three-spined stickleback, he is immediately attacked.

Territoriality and aggressiveness were found to be highly variable. The defender of a given territory becomes more and more ferocious the closer he is to the center of the territory. While patrolling his borders, he may be driven back by an invading, neighboring male stickleback, but as he retreats and gets closer to his nest, his courage builds. Simultaneously, the courage of the intruder seems to sag as he gets farther and farther from his home territory.

The tide of battle makes a dramatic reversal and the previously indomitable invader retreats, hotly pursued by the now victorious defender, but the victor's ardor for pursuit falls off as he chases the intruder into the intruder's home territory while the intruder regains his fighting spirit once on his home court. Again, the tide of battle turns. So, there is a back-and-forth swing of victory and defeat. It is appealing to compare this fish's behavior to human history. Certainly, in athletic competition, the so-called home court advantage is well known.

Animals of the Water's Surface

Water molecules are attracted to one another equally from all sides except at the air-water interface. Here, because there are no water molecules above them, the surface molecules are very strongly attracted to one another and form an elastic film that behaves like a stretched membrane. A number of animals have evolved ways to exploit this film. Some use it as a means of escape; several species of tropical basilisk lizards have adapted to run over the taut surface of standing water.

The river crosser of Mexico and the Jesus Cristo lizard of South America are very fast runners. When they hit water at full speed, they skim over the surface. Although they might be pursued by a larger, heavier predator, these lizards hit the water's surface and skip over it like a thrown stone, while their pursuer breaks through the surface film and bogs down, allowing the lizard to make its escape.

There is one mammal that deserves a Jesus Christo name since it, too, can walk on the water's surface. This animal is a tiny shrew, *Microsorex*. Weighing only a few grams, microsorex has a heartbeat of over 1,500 beats per minute and is a voracious hunter capable of eating more than its weight in a day. It even has a venomous saliva. With all its awesome skills, the shrew is hunted by larger, heavier predators, and it too uses the surface tension film to escape.

Over 300 million years ago, the first animals to adapt to the surface tension film appeared. They were a group of wingless, tiny insects now known as springtails. The surface tension film easily supported their lightweight water-repellent bodies. Like all insects, the springtail has three pairs of legs. On its ventral surface are two unusual organs, a wettable colophore, an organ that adheres to the surface, thus preventing wind from blowing the springtails around, and a special mousetrap-like appendage, the furcula, attached to the tail and folded forward against the animal's abdomen. This spring is held in place by a latch which the insect can unhook on command, allowing the energy stored in the spring to flip the little bug into the air for about 15 body lengths. It is an excellent escape mechanism.

Joint-Legged Denizens of Calm Waters

The air-water interface and the resulting surface tension film brought new environs and opportunities to be exploited by small, lightweight arthropods such as insects and spiders. Some organisms exploited the surface film of calm waters supported by this elastic membrane by walking on it, skating on it, or hanging from it. Some even used it as a launching pad. Other animals utilized air trapped in surface tension film bubbles and attached it to their bodies as underwater scuba gear. Among these scuba divers are a group of fishing spiders that construct underwater nests out of air bubbles. For some air-breathing aquatic insects, the surface tension film posed a problem in that they had to come to the surface and penetrate the film to get air; over time, evolution produced snorkel tubes which allowed the animal to stay submerged while breathing.

Ripple-Detecting Predators
of the Water's Surface

Beetles constitute the largest and most highly advanced group of insects. So it is not surprising that over 1,000 of the known 30,000 species of beetles adapted to aquatic life. In doing so, most of them evolved as air breathers in adulthood. They must regularly replenish their air supply, though some can store air bubbles, whose buoyancy aids in surfacing after they dive. One of the most successful of the aquatic beetles is the jet black, oval, somewhat-flattened, small whirligig. This beetle is seen in schools wheeling at high speed, half in and half out of the surface tension film. Whirligigs propel themselves by a rowing action of their middle and hind legs, while the front legs are raptorial and are held in a forward position to grab their prey. As they paddle forward, they push up an impressive bow wave, creating ripples which extend a considerable distance in front of the insect. At maximum speed, they can cover 16 inches per second, or almost 1 mile per hour, a speed which is rather impressive for such a small aquatic animal. They are frequently seen in aggregations or schools of over 100 individuals, zigzagging and whirling on the surface but never colliding. Most of their information comes from their short antennae that can detect ripples reflected back from an object. This system is so sensitive that they can perceive prey or obstacles even in total darkness. However, whirligigs do use vision, having a pair of compound eyes that are divided so that the dry upper half can scan the air and the surface, while the submerged wettable half can scan the underwater world. In addition, these beetles have evolved chemical defenses against predators. They can, when disturbed, secrete a vile-smelling defensive odor that to some smells like spoiled apples; this is why the whirligig acquired the nickname apple bug.

There are also real bugs (Hemiptera) that frequent the surface tension film. The back swimmers, fierce, half-inch-long, torpedo-shaped bugs, hang upside down by their rear ends from the surface tension film. They use their hair-covered hind legs as oars, and their diving skills are excellent. With their compound eyes, they scan the water for prey while the ripple-sensitive tip of their middle legs monitors the surface for high-speed vibrations, such as those caused when a small insect falls into the water. The back swimmer then moves to its target and attacks from below. With its conelike piercing-sucking beak, it injects toxins and enzymes and digests the innards of its prey. The beak is powerful enough to penetrate human skin. Back swimmers are downright antisocial, solitary hunters. They will attack and eat anything, including their own offspring.

Another group of aquatic predatory bugs are the water scorpions. These large, elongated predators have powerful raptorial front legs and are equipped with a well-developed, long breathing tube, or snorkel, at the end of their abdomens. They are visual hunters but may be able to detect vibrations in the water with sense organs on their abdomen. However, the uncontested top gun of the water bugs is a giant belonging to the genus *Belastoma*. Some of these water bugs are 2½ inches long and are extremely voracious, spectacular hunters whose colloquial names "fish killers" and "toe biters" attest to their skills and weaponry. This bug not only has clawed raptorial forelegs but is also equipped with a long, sharp beak that is kept folded up on its ventral side as it hangs from the surface tension film by the tip of its abdomen. Despite being powerful and well armed, it will often feign death when disturbed. These giant water bugs are also spectacular fliers and are attracted to lights at night, giving rise to another nickname, "electric light bug."

One group of bugs, the water striders, water runners, and water skaters, have become fully adapted to exploit the surface film of calm waters. These small predators skate on the surface film supported by their water-repellent, hair-covered feet and have lower hind limbs and long slender middle legs that act as oars to propel themselves. The short front legs can provide additional support but can also be used in prey capture. The middle legs and hind legs have vibration detectors on their feet that can notice even small potential meals. So, a quadrant of receptors provides enough accurate information to allow the strider to orient itself toward prey and pivot away from potential predators and flee. Furthermore, they are aided by visual information from their large compound eyes, which are also used in hunting. Typically, when it is hunting for subsurface-dwelling prey, the strider will purposefully move toward its prey, unfold its long, sharp proboscis, spear the insect, inject its

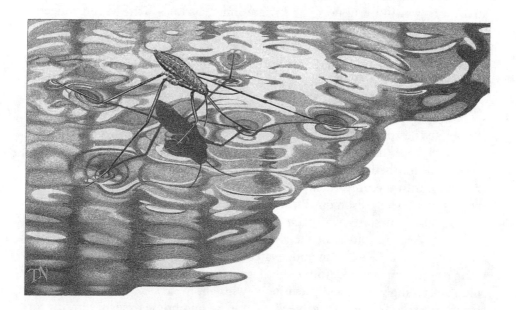

Walking on Water and Detecting Ripples   Some predatory insects, such as this water strider, locate their prey with vibrational sensors that detect ripples on the water's surface.

enzymes, suck it dry, and then discard it without ever having brought the prey to the surface. Striders can not only detect ripples but also generate their own ripples, which serve to communicate information to other striders. We know that some signals are threats to delineate territory or potential mating sites, and other signals indicate a male's desire and female's acceptance.

Water Spiders

Several spiders have evolved to exploit the rich hunting ground under and on the water's surface. To attain fishing capabilities, these spiders have become scuba experts that can carry an air supply with them under the surface. Being air breathers, they must periodically come to the surface to replenish their air supply. In England, there is one wolf spider, *Lycosa purbeckensis,* that is an ocean hunter that lives on beaches. Most fishing wolf spiders, however, are found in freshwater. The pirate spider, *Pirata piratica,* is found along pond edges with its forelegs resting on the water's surface. It can detect tiny vibrations. Insects that fall onto the water's surface set up ripples as they struggle to escape from the tension film, and the spider responds by running across the surface and making the kill. However, the spider is not limited to the surface and is an excellent diver, submerging to escape other predators or to hunt fish considerably larger than itself.

Fisher spiders of the genus *Dolomedes* can not only detect vibrations created by fish swimming near the surface but may lure them within range by tickling the surface with their mouth parts. Wolf spiders are jumpers, and once the target is near enough, *Dolomedes* gracefully skates across the surface buoyed up by air bubbles trapped on its water-repellent leg hairs. Because those air bubbles make it so buoyant, the spider must push off against some solid object or cling to some submerged object to get at its underwater prey, which includes small fish, tadpoles, and salamanders.

While the spider is underwater, the trapped air bubble provides this scuba diver with a continuous, half-hour-long supply of oxygen. Its weapon system is a powerful toothed-fang-bearing structure, the chelicerae. These hollow fangs are driven into the prey; then the spider injects a mixture of poison and digestive enzymes. The poison quickly immobilizes the victim, and the enzymes dissolve the prey's innards, which are then sucked up by the spider.

Other aquatic spiders of the genus *Argyroneta,* found in Asia and Europe, are even more adapted to an underwater hunting role. These swimming spiders bring trapped air bubbles from the surface to an underwater diving bell made of silk. Painstakingly built and provisioned with air, these silken, air-filled tentlike structures serve as a residence, breeding ground, and home base for hunting forays. Being an air breather, *G. Argyroneta* carries its scuba tank, an air-filled globule, attached to hairs on its abdomen. It is a rapid swimmer and has no problem catching and immobilizing aquatic prey five times bigger than itself.

Chapter 3

The Terrestrial Battleground

In the Beginning

During the great Cambrian explosion of new and varied life-forms, there were no terrestrial organisms. The biological big bang occurred on the primitive seas, and the early land mass was a barren and inhospitable moonscape. The continents, as we know them, had yet to drift apart, and, thanks to oceanic photosynthesis by bacteria and blue-green algae, the atmosphere was gradually accumulating oxygen. With the increase in atmospheric oxygen (O_2), the thin upper layers of air, being bombarded by high-energy radiations, began to accumulate a new molecular species of oxygen, O_3 or ozone. Because of its three-dimensional shape, ozone absorbs specific wavelengths of ultraviolet radiation that damage DNA. If one looks at the DNA absorption ultraviolet spectrum and compares that with mutation frequency, it is clear that those wavelengths which cause the greatest biological damage are the very same wavelengths which are absorbed by ozone. Having such a radiation shield was an essential step that allowed organisms to move onto exposed land surfaces.

Another extraordinary event had to occur before animals could invade and exploit the bleak early landscapes: colonization of the land by plants. This was a daunting task which was carried out gradually over the span of many millions of years. In order for plants to move from a watery environment to dry land, they had to carry within their tissues a fluid remnant of their evolutionary past. However, the surface of the land is episodically dry, and during such dry periods, plants would desiccate and die. In order for plants to survive on land, then, one prime requirement was the evolution of an outer waxy covering to minimize water loss. Once that requirement was fulfilled, plants at the land-water margins had the equipment to establish a beachhead into the as yet unexploited terrestrial bonanza.

The Landing of the Invertebrates

Once established on land, the greening of the planet established a myriad of environments that could be exploited by animal life. As was the case with plants, the invasion of land by animals began at the shore lines, beaches, and riverbanks, where freshwater flowed into, and mixed with, the salty seas to create less-salty estuaries. However, for animals to move onto land, they needed a locomotion mechanism sturdy enough to support their weight. Along the shorelines, there were a multitude of creatures with "the right stuff," the arthropods. In the vanguard of arthropods establishing beachheads on land were ancient, six-legged ancestral insects and their eight-legged cousins the arachnids (spiders, scorpions, and mites). According to the indications of these animals' fossils, by the early Devonian period (500,000,000 years ago), invertebrates were moving inland.

The original arthropod invaders were small. Natural selection had provided them with a remarkable external skeleton composed of several layers of armor. The thin outer layer was waxy, therby preventing water loss. The inner layers, composed of protein and sugar polymers, called chitin, were bonded together in such a way as to produce an exceptionally strong, rigid, lightweight protective covering. The exoskeleton also provided multiple points for attachment of muscles. The legs of these creatures, composed of muscle-packed hollow tubes, were jointed, providing them with locomotor equipment that permitted them to run, jump, and climb. Living in a rigid armored case, however, presents a monumental problem for the arthropod. In order to grow, it must shed the old exoskeleton while making a new one. This process is called molting. Molting takes time and energy, and during this period the animal is vulnerable. Fortunately, these disadvantages are outweighed by the advantages provided by their exoskeletons.

Another prerequisite for animals on land was the evolution of equipment that would allow them to breathe gaseous air. (In the water, small animals could breathe through their surfaces.) Larger multicellular animals evolved structures such as gills to extract dissolved oxygen from water and specialized oxygen-carrying blood cells to carry the absorbed oxygen to all their tissues. Later, they evolved a network of tubes and pumps to carry blood to the gills for oxygenation and then to carry the oxygenated blood cells to all their tissues and organs. In order to exploit the land, arthropods had to carry within themselves large, moist surfaces for gas exchange and some mechanisms for pumping air into and out of these structures. The insects solved this problem by evolving a system of gated tubules that branched into smaller and smaller tubules that reached within a few cell lengths of all of their cells. In addition, the arachnids developed internal moist chambers containing many crimped folds of tissue called gill books and lung books.

Another part of life's game plan to live on land was to somehow carry within the organism the remnants of its past. In effect, each organism would contain a carefully balanced internal sea to bathe all the cells and tissues. The internal sea, a slightly salty, slightly alkaline carrier of nutrients and gases, was also a repository for the by-products of metabolism, particularly those containing nitrogen. For optimal function, these internal seas had to be very carefully regulated. Excess waste products such as uric acid and urea had to be gotten rid of, and internal water-salt balance maintained by regulated intake and water-preserving systems. These regulatory structures were the kidneys. Land-dwelling arthropods developed specialized structures, called Malpighian tubules, attached to their hind gut. These structures could dump uric acid for elimination while reabsorbing and conserving water at the same time. The ability to preserve water, even in bone-dry desert environments, allowed spiders, scorpions, and insects to thrive in environments so arid that few others could survive there.

System Maintenance

All organisms are designed with built-in maintenance schedules that ensure maximal function of all parts. If any components of the whole system function below par, this decreases the continued existence of that organism, be it predator or prey. As it turns out, organisms are constantly rebuilding almost all of their parts, recycling molecules whenever possible and dumping breakdown or waste products. This whole replacement process goes on continuously, yet the organism retains the same structure. In such a state of dynamic equilibrium, the genetic program, or blueprint, provides instructions for mandatory replacement of parts.

The instructions inscribed in the double helical molecule of DNA must be exceptionally stable, and maintaining the stability of a complex molecule held together by weak bonds is the first order of evolutionary business. Over the course of the history of life, DNA developed a number of repair mechanisms, molecules that constantly edited the DNA for damage. Once an error was detected, the damaged portion of the program was cut out and the damage was repaired. The higher we go on the evolutionary scale, the greater the DNA repair capacity. As of this writing, we know of more than thirty different DNA repair enzymes in humans. Being nonredundant, DNA must have such a repair system, but the products produced by DNA, namely the proteins, number in the millions and such redundancy can tolerate considerable damage without significantly impairing function.

Redundancy works at a number of levels. Many organs are present in pairs. Should one of the pairs be lost, the function doesn't really alter, and the loss is compensated for by the growth of the remaining structure. As a matter of fact, most replacement continues automatically by feedback signals. For example, if a wound causes blood loss, the kidneys detect the decrease and release a chemical messenger that stimulates the bone marrow to increase the production of new red blood cells.

Other than normally scheduled maintenance, many organisms have a multitude of systems that repair injury. Tears in blood vessels automatically trigger the clotting mechanism, a cascade of some 13 proteins to form clots and plug the tear. Tissue damage activates controlled proliferation by nearby cells and the formation of strong scar tissue. All organisms, both predators and prey, become battle scarred over time, and natural selection favors those organisms that can precisely repair damage and restore essential parts. In some animals, this restoration can and does involve significant loss of body parts. Starfish, for example, if cut in half, can regenerate the two halves into new starfish. Crabs and lobsters can, at will, lose appendages grasped by a predator and, in time, regenerate the lost appendage. This ability to give up a part to a predator, rather than lose the whole organism, is a successful escape strategy. As noted in Chapter 1, a number of lizards, when pursued, shed a piece of their tail which, although it is no longer attached, continues to wiggle and distracts the predator from the larger meal. For the predator, in such cases, a piece of tail is better than nothing at all. Not all animals can manage such spectacular feats of regeneration, but most can repair bone, grow new blood vessels, and even repair damaged nerve circuits.

In addition to replacement and damage control, many animals can produce structural changes in response to environmental demands. The more a muscle is used, the bigger it gets. It adds contractile proteins, elastic tendon, and increased blood flow; and even the bones attached to the muscle increase mass and mineralization. This process is called work hypertrophy. For example, bodybuilders, such as Arnold Schwarzenegger, dramatically increase their muscle mass by heavy exercise. Conversely, if bone and muscle isn't used, it shrinks, or atrophies. In some forms of competition, such as the battle for resources (space and mates), increased musculature, strength, and speed enhance fitness and produce winners, but not always. As we go up the evolutionary ladder, the ability to learn, to deceive, to use guile, and to create cooperative learning strategies are more important in producing fitness.

Terrestrial Warriors without Backbones

The vast variety of invertebrates and their varied offensive and defensive weaponry could fill several volumes. In this chapter, we consider only a small number of some of the more-fascinating land-dwelling invertebrates in detail. The two most successful phyla are the insects and the arachnids. It is worth noting that the insects have developed a degree of mandatory social cooperation exceeded by no other group of animals.

Most organisms, particularly organisms that are predators, live as solitary individuals programmed to preserve themselves and, on occasion, to mate with a member of their own species. Other animals exhibit episodic aggregations at some stage of their life cycle and, during these times, communicate and cooperate with each other for the greater good of the species. Still other groups of animals have evolved to a truly social status in which individuals are totally dependent on the colony. The most advanced societies are seen among the insects, the Hymenoptera (ants, wasps, and social bees) and the Isoptera (termites). In their societies, generations of parents and offspring live communally, usually in a central structure, or nest, constructed by members of the colony. At the center of these societies is a queen, the source of all members of the colony, and a number of different types of workers whose programs include maintenance of the queen, her eggs, the pupae, and the larvae, as well as food gathering and defense of the colony.

Within the colony are individuals whose structure and behavior are highly specialized for their particular role in the preservation of the whole colony. Cooperation has proven to be a robust strategy for survival. But inherent in the concept of cooperation among large numbers of diverse individuals is that each member of the group can be relied on to fulfill his designated task. In social insects, the specialized behaviors are inherited (one might say hardwired) into the circuits of the animal's brain. There is no flexibility, no consciousness, no remorse, no guilt; there are no questions, no risk or cost-benefit decisions, and no ambiguities. These circuits produce mandatory stereotypic responses to a stimulus, usually an olfactory stimulus. The reliability of such responses ensures survival of the colony. So, when a sentry guarding a nest entrance detects an approaching predator and releases his alarm signal, all soldiers in the vicinity are recruited to defend the colony, even though it may cost them their lives.

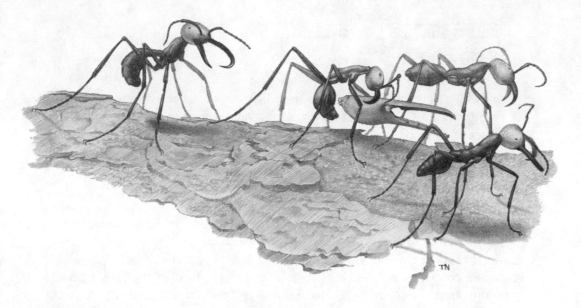

Unstoppable Hordes   Marching in efficient military formations, army ants sweep all living things before them, killing and dismembering anything in their path. Even six-ton elephants are attacked and will flee the massed approach of these tiny warriors.

The Army Ants

The late director of the National Zoological Park in Washington, D.C., Dr. W. Mann, while studying army ants, characterized them as being "like the armies of Ghengis Khan traveling in countless hordes, marching in efficient military formations and sweeping all living things before them, leaving a wake of death behind them." Such colorful anthropomorphic descriptions ascribe to these terrestrial predators a ruthless malevolence, equating their behavior to the dark amoral side of human nature. But such description ignores the biology underlying these social animals. Their behavior cannot be a conscious behavior, because they have no central mental processing about right or wrong, nor do they have compassion or cultural restraints. What they do have is a relatively simple repertoire of behaviors that ensure they will survive as a species, implanted into their relatively simple neural circuits.

Army ants, also called driver or legionary ants, live in most temperate climates. The largest of them and their armies are found in the tropics, where they live in colonies of hundreds of thousands to millions of individuals.

These blind, vast armies can overcome all obstacles including mechanical barriers and even water. They have scouts; small soldiers; large, powerful brigadier warriors; and even engineers that build bridges with their own bodies. Such complex activities are highly organized, whether carried out during the periodic movement of the whole colony, during temporary bivouacs, or during their food-foraging raids. During these episodic foraging exercises, they kill and dismember anything in their path. Army ants are carnivores; although their main prey are often arthropods, there are many well documented examples of a swarm's killing large animals in order to provide the tremendous amount of food needed to provide sustenance for the colony.

In order to coordinate their behaviors, these ants communicate via their sense of smell. The chemical signals called pheromones dictate a number of specific behaviors such as trail marking, alarm warnings, recruitment signals that call up the reinforcements needed to overwhelm a prey, and colony identity signals secreted by the queen. The smallest workers of the colony stay with the queen and service the larvae and pupae, while the intermediate-sized workers, who are the most abundant individuals, compose most of the foraging column, following the scout's trail markers. On the column's flanks are the largest ants, the formidable soldiers or brigadiers. Each brigadier has a pair of large, very sharp, ice-tong-like curved jaws and a potent venom-injecting stinger on its tail end. A foraging column may extend a hundred meters or more, and within the column, the traffic flow is in two directions. While the soldiers move forward, intermediate-sized worker-soldiers carry dismembered bits of prey back to feed the rest of the colony.

In our home in the Kenyan bush, we had a huge colony of army ants and my young children quickly learned to always look down to make sure they didn't step on a foraging column because the results could be very painful. I remember having to pry open the jaws of a couple of brigadiers that had latched onto my son. The local people said you could use the tong-like jaws as wound clips or sutures to close a cut. They would pick up a brigadier, hold it against the cut until the jaws snapped shut, and then, with a twist, separate the head from the rest of the body, leaving the wound firmly closed. During bivouac nesting or on encountering a water obstacle, workers use their own bodies to form strands held together by interlocking jaws.

During one ant migration, the whole column came up the wall of our house and flowed into the living room through a screenless window. I used five cans of pyrethrin insecticide spray, until the floor was literally covered with dead ants, but they kept coming. The gardener then came running up with a huge container of propane gas attached to a nozzle and, using a jet of flame, burned up about 100 feet of the column. Of course, that started a brush fire which eventually burned out, but the emotional impact of an army ant invasion is sufficient to evoke dumb heroic measures.

The Fire Ants

Fire ants of the genus *Solenopsis,* though tiny, are among the most insatiable, aggressive, and terribly fecund ant species. They live in huge underground supercolonies which may house more than 500 queens. Each of the millions of workers is equipped with strong pincer-like mandibles and a potent stinger, a needle-sharp, barbless hypodermic, through which poison is injected. Introduced into the United States in 1918, fire ants have spread with astounding swiftness throughout most of the southern states. They have infested over 25 million acres, and a single acre can be home to more than 20 million ants, or about 500 individuals per square foot. Fire ants send hordes of foragers to search for food, which is ingested and then carried back to the nest, where it is regurgitated to feed their nest-bound, hungry cohorts. Guidance back to the nest, which may be more than 100 feet away, is achieved by trail-marking pheromones. There is some evidence suggesting that these ants are also capable of solar or lunar navigation.

Once they latch onto a prey or intruder, they can sting repeatedly with their barbless stingers. Their venom is a lot less toxic than that of some other ants, such as the Costa Rican ant called "bullet" (because when it stings, you feel as if you'd been shot) or the giant red Australian bulldog ant, whose sting is so toxic that 30 stings can kill an adult human. However, despite the mild toxicity of their venom, the fire ants are so numerous and so aggressive that they can deliver thousands of stings to an intruder. It has been estimated that 5 million people per year have been stung by them and that 40,000 to 60,000 of these victims required visits to the emergency room. Despite massive efforts to control them by using the most-advanced technology, humans have been unable to deter their unrelenting advance. Chalk one up for the ants.

While the spread of imported fire ants has so far been impervious to human enemies, these creatures are not immune to attack by a tiny, airborne parasitic fly, *Pseudacteon solenopsidis,* that hovers in front of one and strikes with its sharp ovipositor, injecting its eggs. The eggs develop into larvae and eat the infected fire ant.

The Weaver Ants

The pinnacle of ant evolution exists in the fierce weaver ant colonies of Australia. Not only do these large-eyed, tree-dwelling ants have a sophisticated

chemical language that includes trail-marking, recruiting, and alarm pheromones, but they also send signals by their postures and body language. They
may have as many as 50 different signals. Powerful, pointed, serrated jaws serve
as their armament, and their bite is augmented by glands that squirt formic
acid and, possibly, other toxic chemicals. These animals are very territorial, setting up boundaries for their colony that may contain more than a dozen trees.
Members of their own colony who have "identify friend" odors are allowed to
enter, but any intruder entering their domain is attacked. When a guard weaver
identifies an entering animal as a foe, the guard releases a recruiting
pheromone, a call to battle stations. Soon, hundreds or even thousands of
defenders race from their nests to the battleground and launch a mass attack
on the intruder, biting and holding him captive with their jaws and spraying
him with formic acid until he is subdued. Unlike army ants which send out foraging columns, harvesters limit their aggression to their own territories. Their
attacks are so effective that there is no need to leave their home range.

Recruiting Ants as Mercenaries

Only ants, termites, and humans employ a specialized soldier caste to defend
the society as a whole. Although the cost of maintaining an army is high, it has
proven to be a cost-effective overall strategy. However, there are some mutual
interactions in which an otherwise defenseless organism of one species recruits
ants by providing sustenance and lodging to fierce warrior mercenaries. One
example of such mutualism is evident in the interaction between the swollen
thorn acacia trees and their resident warrior mercenaries. Unable to run away,
plants are munched on by a great variety of herbivore predators. Over the
course of time, most plants have evolved a variety of defenses, mainly chemical,
to counter the hunters. (We consider these remarkable defenses in detail in
Chapter 7, "How Plants and Fungi Make War.") Some especially delectable
plants, such as the swollen thorn acacia, lack chemical defenses and so have utilized hired guns to protect themselves. The resident defenders get sustenance
from the syrup-like, protein-rich secretions of the acacia. Resident barracks for
the queen and workers are provided in hollowed-out thorns. In return for this
largesse, the resident warriors, detecting any invading plant muncher, sound
the alarm and rush to the attack, driving off the intruder with potent stings.
The mercenaries also protect their host plant by clipping off sprouts of competing plants, which if unimpeded, would overgrow and shade out the swollen
thorn acacia. Using their powerful, sharp mandibles as pruning shears, the ants
clear a circular area of up to 5 feet around the acacia. Over the course of time,
this relationship has evolved so that neither tree nor ants can survive without
the other.

Ant Navigation in the Desert

Navigation involves getting from point *A,* the home base, to a desired destination, point *B,* and back again. Most species of ants achieve this by leaving a scent trail with marking pheromones. The scout ants mark the trail first. Then other ants, following the marker left by the scouts, add their own markers to reinforce the scent trail which would otherwise dissipate in a matter of minutes. However, there is one ant that uses an amazing alternative navigational system. This ant, *Calaglyphis bombycina,* is one of the fastest-running insects, 36 inches per second, and has a foraging range of up to 200 yards. The group leaves its underground nest in the heat of high noon, when the air temperature reaches 116°F in the Sahara desert. After a hunt, they unerringly find their way back to their home base in a terrain devoid of landmarks.

The barren desert, like the vast expanse of the ocean, poses a real problem for navigators. During World War II, thousands of members of air crews lost their lives in the desert due to directional errors despite an array of modern aids to navigation. One aid was a sextant and a set of tables that used the positions of the Sun and stars as guiding lights. But solar navigation had very limited effectiveness, even when used by a trained, big-brained human. Ants' brains are minuscule, and yet they can process information supplied by three tiny photoreceptors in their forehead, the ocelli, to recognize the polarized light pattern in the sky which helps them find their nest. These same ants also have a pair of large compound eyes, similar to those of other insects, but if these are painted over, the ant can still find its way home, using its tiny ocelli. As a former aerial navigator who recalls how often he was lost during World War II, I find these extraordinary feats of navigation by such tiny-brained desert ants truly humbling and thoroughly amazing.

In the intense heat of the Sahara, ants are at minimal risk from predators, since most predators cannot tolerate the high temperatures. But there are times when the desert floor gets too hot even for the ants. To escape the heat, they climb into available plants, making themselves vulnerable to predation by heat-tolerant lizards of the genus *Acanthodactylus,* whose burrows are often located near the ants' nest. When sitting exposed on the stem, the ants are quickly gobbled up by the watchful lizards.

Ant Hunters

Despite their extraordinary chemical defenses, armor, and programmed tactics, ants still represent an abundant and ubiquitous source of highly

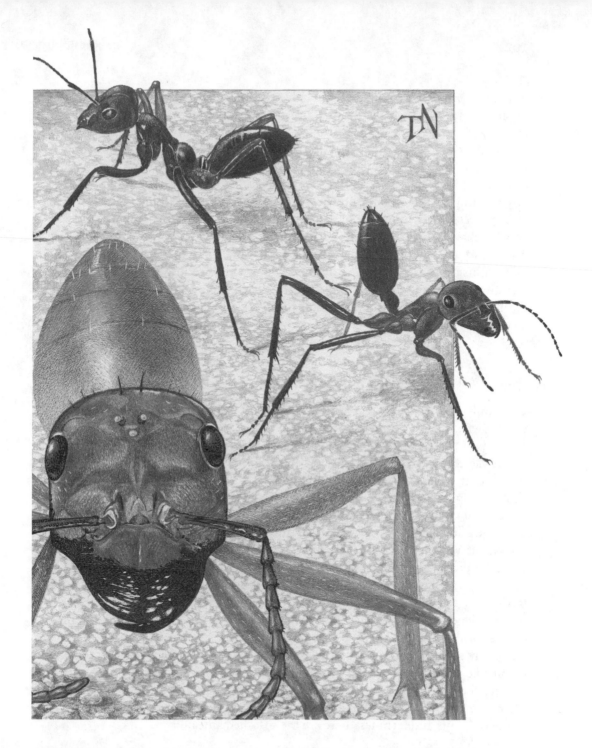

Getting Home by the Light of Day   Most ants follow scent trails to find their way home, but in the trackless heat of the Sahara desert scent markers rapidly dissipate. So, this desert dweller—*Calaglyphis bombycina*—unerringly returns to its nest after a foraging expedition by using polarized rays of sunlight to guide it homeward.

nutritious, readily digestible food. Indeed they are a resource exploited by other ants and by a wide variety of invertebrate and vertebrate predators. The tactics and strategies of some of these predators are quite remarkable. True to its name, the "highwayman beetle" (genus *Amphotis*) approaches foraging black wood ants and stimulates the ants to regurgitate food. When the ruse is discovered, the beetle retreats into her ant-resistant armored shell. Other beetles, working in packs, pounce on ants and immobilize and eat them. They have also perfected a chemical defense system, a secretion that turns off the ants' attack behavior, a sort of chemical pacifier or tranquilizer. However, the most insidious tactic of beetles that prey on ants is that of a short-winged beetle. Like an insidious fifth column, this beetle gains entrance into the ants' nest by secreting a pheromone that induces the worker ants to feed it and even care for the beetle larvae. Then, adding insult to injury, the beetle larvae eat the ant larvae.

The Ant Lion

The adult neuroptera, insects known as ant lions, are fragile, nonpredatory weak fliers. Their larvae, however, are voracious predators whose food of choice is ants. There are more than 600 species of ant lions, and in all of them, the larvae construct conelike unstable pits in sandy soil, and then burrow into the sand at the bottom and wait. An ant wandering into the pit has difficulty climbing up the loose, sandy sides of the pit, and its struggles bring the ant lion out of hiding. First, using its mouth parts, the ant lion throws more sand at the struggling ant. Then, as the ant slides to the bottom, the ant lion pounces on it, impales it with its sickle-shaped mouth parts, pulls it into the sand, injects powerful digestive enzymes, and sucks out the prey's digested innards.

Many vertebrates eat ants and termites. Some, such as the toads, lizards, birds, chimpanzees, and even humans, are opportunistic feeders. Chimpanzees, using a slender stem as a tool, probe into termites' nests and pull out the stem covered with termites. After shedding their wings, termite aylates are numerous and easily gathered. Humans collect them and either eat them raw or fry them in oil like miniature popcorn shrimp. They actually taste quite good. Some novelty food stores even offer chocolate covered ants as a gourmet specialty. There are also a number of mammalian specialist species whose primary food resource is ants and termites; these are the so-called anteaters.

The Termites

Some 150 million years ago, a roachlike ancestor gave rise to the termites, one of the most successful forms of life on earth, particularly in the tropics. Termites are the most primitive of the social insects and are organized in rigidly structured societies. Each colony contains several distinct forms (soldiers for defense, workers, kings, and queens), each carrying out a specific function within the society. All termites are blind, so most communication within the colony is accomplished through chemical signals (smell and taste). One characteristic of colonial life is the sharing of food among individuals; for termites, this procedure includes the transfer of wood-digesting protozoa and bacteria that inhabit the termites' bowel. Termites have chewing equipment and eat wood. Most other animals cannot digest wood, but termites' bowel-dwelling microorganisms can break down cellulose into its component parts, which are simple sugars.

Evolution

A primal termite knocked on wood
Tasted it, and found it good
That is why your cousin May
Fell through the parlor floor today.

OGDEN NASH

Termites are so numerous that they, along with ants, make up 75 percent of the planet's animal biomass. At this moment in time, there are more living termites and ants than all the people who ever lived. This being the case, it is not surprising that there is a great variety of predators adapted to exploit this abundant nutrient source. In order to counter predators, termites not only live in tunneled nests but also have a full-time army equipped with an amazingly diverse arsenal of mechanical and chemical weapon delivery systems.

The soldiers of the termite armies are sterile, blind, soft-bodied, six-legged creatures with highly specialized heads. Their only function is to defend the colony from birth to death. These walking weapon systems are programmed to blindly carry out their defensive role. Some species of soldier termites are armed with powerful biting and snapping saber-like jaws that can penetrate the protective shields of attacking predators. These mechanical armor-piercing weapons were supplemented later in evolution by a bizarre chemical weapon arsenal composed of repellents, toxins, sticky glues, and contact poisons. The tactics for delivering those chemical defenses vary from species to species. Some species, after wounding an intruder, secrete a potent anticoagulent which is painted or injected onto the wound with a modified brushlike lip. In other species, the *Anoplotermes*, the

Glue Guns and Pinching Jaws   The soldier castes of different termite societies defend their nests with a variety of weapon systems. The *Nasutitermes* on the left squirts a sticky, malodorous, distasteful goo at its attacker, while the soldier termite on the right uses its powerful pointed pincers to defend the colony.

soldier caste defending the colony rushes to attack any intruder in a kamikaze-type charge; the soldiers then constrict a powerful ring of muscle around their abdomens and explode bathing the attacker with their toxic gut contents.

The more recently evolved *Nasutitermes* soldiers have bulbous enlarged heads equipped with an elongated, glue-squirting, bazooka-like snout atop their foreheads. The toxins make up over 10 percent of these soldier's dry weight, and some species not only store toxin in their heads, but also have spare ammunition depots in their abdomens. In these termite soldiers, 35 percent of their weight is dedicated to chemical weaponry, including glue guns. The glue squirters entangle their attackers with a barrage of malodorous, distasteful, viscous goo that repels even giant anteaters. After facing the determined and coordinated attack of hordes of defenders, attackers are quickly driven off. These same chemicals are even more effective in protecting termite colonies from their main predators, the ants.

Festung Termitica

Many species of termites construct fortress-like nests made of soil or sand grains glued together by the termites' own secretions and baked hard by the tropical sun. Some termite mounds in Africa tower over 15 feet high and are interlaced with temperature-controlled interior chambers that house the queen and her brood. The tunnels connecting these chambers vent at the top. When wind blows over these openings, it pulls the air in the tunnels out, creating an airflow that carries the heat generated by the metabolism of a million or more termites out of the brood chambers. Other species of tropical termites construct similar sand grain nests in trees, where they are safe from most marauders digging for termite snacks.

The Anteaters

The portals to both the ants' terrestrial and arboreal fortresses are guarded by soldier members of the colony, whose alarm pheromones can call up large numbers of reserves to ward off any would-be invader. However, for every defense, predators have evolved a way of subverting it. For example, over time, one species of tropical assassin bug has evolved seasonal tactics to overcome the defenses of arboreal termites. The first step is a stealth technique that prevents detection by the sentries. The assassin bug covers itself with bits and pieces of nest debris, making itself invisible. The bug then stations itself by a nest opening. After it snatches a termite and sucks it dry, it then dangles the carcass of the termite over the opening. Termites and ants recycle the nutrients of their own dead, and one can frequently see them carrying deceased nest mates home for these purposes. A termite, seeing this apparently dead fellow, will grab it, only to be gradually pulled out of the nest by the assassin bug and devoured. This lure-style fishing for termites by the patient assassin bug may continue for hours.

There are other animals that have evolved specialized hunting tactics that exclusively use ants and termites as food. There are the anteaters. There are 22 species of these mammals, two of the most primitive of which are in Australia:

An Ancient Anteater   A great variety of mammals have specialized in hunting ants and termites. Among the most ancient of the anteaters is the small, spiny echidna, an egg layer that is related to the duck-billed platypus. It is well protected by its spiny armor and can tear its way into a nest with its powerful, clawed front paws.

the Australian numbat, or banded anteater, found in the eucalyptus forests, a marsupial that evolved its ant eating in isolation from the world of placental mammals, and the tiny spiny anteater, an 11-ounce egg-laying mammal related to the duck-billed platypus. Both species developed adaptations that made them capable of overcoming termite and ant defenses.

In both the New World and the Old World, mammals that hunted ants evolved. The squirrel-sized silky anteater has exploited ants in arboreal habitats, by acquiring a prehensile tail and padded, grasping hind limbs so that it could free its forelimbs armed with razor-sharp claws to rip open nests. In a single night of foraging a single silky can consume 10,000 ants. There are two species of tamanduas, long-nosed olfactory hunters that locate their ant prey by both smell and sound. Their weapon system is a thick, prehensile tongue covered with sticky saliva that allows them to lap up many ants in one lick. However, they are not immune to the defensive weaponry of stinging ants or the irritant chemical defenses of termites. For this reason they tend to feed rapidly and retreat when the defending reinforcements arrive.

The largest of the New World anteaters are the 60-pound giant anteaters, furry, long-nosed terrestrial hunters that specialize in eating carpenter ants. Their tactic is to use their long, sticky tongues to garner the rich harvest available at nest openings. Once the ants are alerted to the threat, however, they retreat to safety in inaccessible parts of the nest. So, to obtain a sufficient supply of nutrients the giant anteater must forage over a considerable distance.

The largest animal that preys exclusively on ants is the aardvark, which derives its name from Afrikans (in which *Aard* means "earth" and *vard* means "pig"). This 6-foot-long, 150-pound termite eater is indeed piglike in appearance, having a sparse covering of hair on its thick, insect-resistant skin and a long-nostriled snout. A nocturnal hunter, it uses smell as its primary means of locating prey. To that end, it has an extensive collection of olfactory receptors, possibly more than any other mammal; in fact, a large portion of its brain is dedicated to receiving and processing olfactory information. When feeding, it can close its external nasal openings, which are also protected by a ring of thick bristles. The aardvark is a superb digger, and its powerfully muscled legs are equipped with massive, stout claws which allow it to burrow its way through the rock-hard, sun-baked walls of the termites' towering nest and gain access to the termites crowded into the inner chambers. When digging for termites, the aardvark props itself on its thick, 2-foot-long tail and hind limbs and uses its forelimbs to penetrate the mound.

Digging also gives the aardvark a daytime safe haven; it digs a labyrinthine tunnel system, 20 feet long. The aardvark is capable of high-speed excavation, using its forelegs to dig and its hind legs to throw the dirt backward. With its long, supersensitive ears, it can detect a predator and burrow out of sight before the predator gets close enough to attack. Should the predator go into the burrow, the aardvark's claws are soon put to use as potent defensive weapons.

The primary eating apparatus of the aardvark and of many other anteaters is a protrusible muscular tongue. As it has been modified for harvesting small prey, it is covered with a copious supply of sticky saliva which is provided by outsized salivary glands. The aardvark's tongue is 18 inches long and can flick in and out 120 times a minute, carrying both sand and termite into its mouth. Aardvarks' mouths lack incisor and canine teeth; located at the back of the cheeks, the flattened, rootless molars, devoid of enamel, crush the ingested termites. Given the abrasive effect of ingested sand, almost 50 percent of the stomach's contents is sand. The molars are continuously being replaced. Although the aardvark is apparently evolutionarily at a dead end, it will survive as long as there are termites.

Also found in southern Africa is the aardwolf ("earth wolf"). The hyena-like appearance of this animal is a form of mimicry that discourages carnivorous predators. The aardwolf locates its prey by smell and an acute sense of hearing. It is not a digger but instead scrapes away lightly covered termite foraging areas and feeds on the exposed termites. When seasonal rains end, and the winged reproductive ant and termite aylates emerge by the hundreds of thousands to seek mates, the aardwolves, along with a great many other animals, become opportunistic feeders, gobbling up the plump, unarmored, defenseless aylates without fear of evoking counterattacks by the soldier castes. But the aardwolf, like many other ant-eating mammals, is not totally immune to ant and termite defenses. Consequently, it feeds for only several seconds before retreating. The end result of such cropping feeding is that nests and colonies, though depleted, recover and can be exploited again another day.

A number of anteaters specialize, feeding only on ants and ignoring the more aggressive and potent termite defense systems. In Australia, there is a 6-inch-long grotesque-looking, prickly, ant-eating lizard called the moloch or thorny devil. Its name, moloch, is derived from a Cannanite god of the Old Testament who demanded the sacrifice of children. Its sharp-thorned armor protects it from predators as it sits by a trail of ants, flicking up 30 to 45 insects per minute with its sticky tongue. In the course of a single feeding, it can consume a few thousand ants, each of which is crushed by serrated teeth at the back of its jaws. The molochs are highly selective in their diet, assiduously avoiding those ant species equipped with a potent stinging apparatus.

The Arachnids—Stingers, Pincers, and Webs

Four hundred million years ago, in the ancient Cambrian seas, other groups of arthropods were preparing to come ashore and take up terrestrial existence. Although all of the ancestors of today's spiders and scorpions are now extinct, in their time they were awesome predators. One ancestor of the spider, a trigonobatid, was as large as a big lobster and equipped with massive jaws. Even so, it was dwarfed by the fearsome sea scorpions, the mighty Eurypetids. Some of these extinct giants, the 6- to 9-foot Slimonia, haunted the estuaries and rivers of the Silurian and Devonian periods, lying in wait in the shallows. Scanning for prey with eyes on top of its head, these stealthy ambush hunters patiently waited for prey to come within range. Then, with a sudden lunge, the open jaws of their pincer-like chelicera snapped shut and dinner was secured. Some of the water scorpions also had a long spike at their rear ends which was possibly a harbinger of the stinger to come later. Gradually, these aquatic eurypetids overcame the transition from saltwater to freshwater, grubbing their way upstream and then onto the muddy shores. Having established this beachhead, they would eventually be able to tap into resources that had hitherto not been exploited.

Nocturnal Predators with Lethal Tails   Scorpions are both ambush hunters and stalkers. A scorpion will eat anything it can subdue with its powerful pincers (usually used for grasping prey) and potently venomous stinger, mounted at its hindmost abdominal segment. Its venom is a mix of nerve and muscle poisons that acts quickly and has been responsible for thousands of human deaths.

By the Carboniferous period, when the atmospheric oxygen levels reached their peak, the landward march had begun. The first terrestrial scorpions were well underway, and they and other terrestrial arthropods independently evolved an arsenal of diverse chemical weapons. Some also reached giant proportions including a archaic 3-foot-long scorpion, which, in form, closely resembled today's creature of the same name.

Today, there are 1,500 species of known scorpions and probably another 1,000 species that have yet to be classified. They are remarkable in their ability to adapt to most environments ranging from the hottest, driest deserts to the greatest heights of the Himalaya Mountains. They can tolerate heat that would kill most animals, turn off their metabolism, remain under water for two days, be supercooled below freezing, and be kept foodless for a year and yet still survive. They are true wunderkinder, whose opportunities for survival in adverse conditions are wider than those of any other animal.

All scorpions are predators that will eat anything they can subdue, including mates and even their own offspring. Although they are both active stalkers and ambush hunters, they spend over 90 percent of their lives in the solitary confines of their burrows. However, if the species is to be preserved, they must come out to hunt and to mate. Like everything else about scorpions, their mating ritual is bizarre. Females announce their sexual readiness by leaving a perfumed trail of species-specific pheromones. When the scent is detected by a male on the hunt, he forgets about food, and his libido kicks into high gear. At this time, the amorous male goes into a lively dance, a sort of spastic rock and roll first described in detail by the great French naturalist, Jean-Henri Fabre. Today this movement is referred to as "juddering." The rhythm of the male's stomping is coded and sets up vibrations in the sand that are decrypted by females of his own species. Using codes is not without risk, and the use of deception, in which false signals have been sent to other species by lightning bugs, is well documented. It has been suggested that some larger scorpions learn the vibrational cryptogram of smaller species of scorpions and judder their code to lure these females out of their burrows and then eat them.

Vibrational sensing not only is used in mating but also is the primary prey detection system used by these nocturnal hunters, afflicted as they are with poor eyesight, poor hearing, and minimal olfactory sensing. When a potential prey moves, either on the surface or below the surface, it sets up vibrational waves that the scorpion can feel. When the ground shakes under his feet, he awakens (yes, scorpions do sleep) and becomes behaviorally alert. His pedipalps, modified mouth parts equipped with powerful grasping pincers, are extended and opened as the scorpion moves slowly forward stalking its prey. Each advance is short, followed by a pause during which the scorpion orients itself toward the source of the vibration. Using two types of sensors on its eight feet, the scorpion can analyze target direction with great accuracy.

Burrowed prey are dug up, and surface prey grasped with the pedipalps; then the scorpion delivers the coup de grâce with its lethal stinger, located at the bulbous end of its segmented tail. Within the bulb are two large venom glands surrounded by smooth muscle. The glands' toxic secretions are carried via ducts to a sharp-tipped, curved, hollow, barbed stinger which is driven into weak points of the prey's armored exoskeleton. The venom, a mix of toxins, is mainly effective on nerves and muscle and is roughly as potent as cobra venom. Two of the scorpions known to be lethal to humans are the large black *Androctonus australis* of the North African desert, which was responsible for the deaths of many soldiers during World War II, and the *Centroides* of the Mojave and Mexican deserts. The *Centroides,* nicknamed the Durango scorpion a few decades ago, was responsible for over 4,000 deaths per year in Mexico. It became such a problem that the Mexican government launched a scorpion eradication program that worked very well. The hunt for scorpions was aided by the use of ultraviolet light, which causes scorpions to glow in the dark. Today a potent antivenin is available and scorpion-caused deaths are rare. However, the human psyche still views scorpions and their cousins, the spiders, as nightmares incarnate. I once saw my 6-foot tall, muscular, athletic son flee in abject terror from a house spider hanging over his bed. He outweighed the crushable little arachnid a million to one, yet there was something about this eight-legged little predator that evoked in him irrational fear, as it does in many people. This fear was exploited recently by Hollywood in the movie *Arachnophobia*.

While the notorious stinging scorpions attract the most attention, there are others that are equally fascinating. The so-called wind scorpions, or Solopugids, got their name because they are so fast, as in racing with the wind. Not only do they have enormous chilcera, but they also have other mouth parts that are leglike in appearance and have adhesive organs at their tips. These ravenous hunters are capable of capturing and eating even small vertebrates. The slow-moving whip scorpions, the Uropygia, exhibit a wide range in size, with the largest reaching a length of almost 3 inches. Like other scorpions, the whip scorpion has pincers at the head end and a whiplike structure at the tail end. On either side of its anus it has a pair of glands, the contents of which it can spray to defend itself. The main component of the spray is acetic acid, or vinegar, which is a potent irritant on its own, but there is also 5 percent caprylic acid, a fat solvent, that penetrates the waxy outer covering of arthropods, allowing the acetic acid to do its job. Because of the conspicuous odor of their defensive secretion, whip scorpions are also called "vinegaroons."

Other Arthropod Chemical Weapons

During the Carboniferous period, the moist floor of the rain forest teemed with some rather strange animals whose modern-day descendants have continued their landward march to new habitats. One ancient group, the legged but wormlike onycophorans, are literally living fossils whose 70 living species are structurally the same as those of Cambrian fossils. Aside from having a central position in the study of invertebrate evolution, they are of particular interest because of a unique weapon system that they evolved to defend themselves. On their heads, they have a pair of forward-pointing squirt guns called oral papillae. When threatened, they shoot twin streams of a viscous fluid at the would-be predator. They can hit their enemy from a distance of 50 centimeters, and the fluid, when exposed to air, rapidly hardens and entangles the intruder. Distant relatives of the onycophorans, are the myriopoda, which include the millipedes, or thousand-leggers, and the centipedes, or hundred-leggers.

There are over 3,000 species of centipedes, including the medium-sized, long-legged house scorpion. The house scorpion gets rid of household pests, while evoking a total lack of appreciation from the humans who derive benefit from good works. Only one centipede is dangerous to humans, namely the giant *Scolopendra gigantea,* which has a large pair of sharp, poisonous claws on either side of its mouth. However, its venom is not particularly lethal to humans. To defend itself from other predators, this centipede has repugnatorial glands whose secretions are, as the name implies, repugnant.

The Diplopoda, with over 7,500 species, are slow moving, heavily armored, and segmented. Each segment of these creatures bears two pairs of short legs. A large tropical specimen, 8 to 10 inches long, when disturbed, rolls itself into a ball the size of a golf ball. During the Carboniferous period, however, there were 6-foot-long giant millipedes that were literally a marching army of legs. Balling up is not the only millipede defense. They are also well armed with chemical defenses, including phenols, aldehydes, and quinones, whose effects we consider in Chapter 7, "How Plants and Fungi Make War." In addition, they harbor a very lethal poison, hydrogen cyanide. The cyanide is secreted by double glands, making it a sort of binary chemical weapon. Some millipedes ooze the cyanide, while others can spray it up to 30 centimeters. Cyanide blocks oxygen utilization in all cells and is one of the most potent metabolic poisons known to man.

Recently, a new millipede chemical weapon was discovered, a kind of tranquilizer similar to the sedative quaalude. When feeding on the millipede, the large wolf spider ingests this compound and within an hour becomes immobilized for several days, during which time it is very vulnerable to predation. Obviously, the ingested substance doesn't protect the eaten millipede, but it may protect the millipedes' gene pool, in that the stoned spider may become conditioned to avoid eating other related millipede.

The most advanced defensive chemical weapon used by arthropods is found in a small group of beetles known as bombardier beetles. I was introduced to them in Africa, when my four-year-old daughter inadvertently became the recipient of a bombardier's irritant spray. She ran screaming into the house, rubbing her eye. The whole region around her eye was terribly inflamed. We rushed her to the emergency room of Nairobi Hospital, where the attending physician explained she had "Nairobi eye," an inflammation caused by a hot, irritating spray produced by bombardier beetles.

These half-inch-long beetles, common in Kenya and other parts of the world, get their name from the audible "pop" they produce when they squirt out their boiling hot, highly irritating munitions from a nozzle at the tip of their flexible abdomens. The beetles, whose marksmanship is exceptional, can aim this nozzle precisely in the direction of an attacker. Their binary weapon system consists of two glands in the abdomen. Each gland has a thin-walled chamber containing hydrogen peroxide and phenols, and a thick-walled chamber containing peroxidases, which are enzymes that can accelerate the rate of hydrogen peroxide breakdown over a million times. When disturbed, a bombardier, on command, can open the valve between the chambers. As the peroxide and phenols react with the enzymes, an enormous amount of heat is gener-

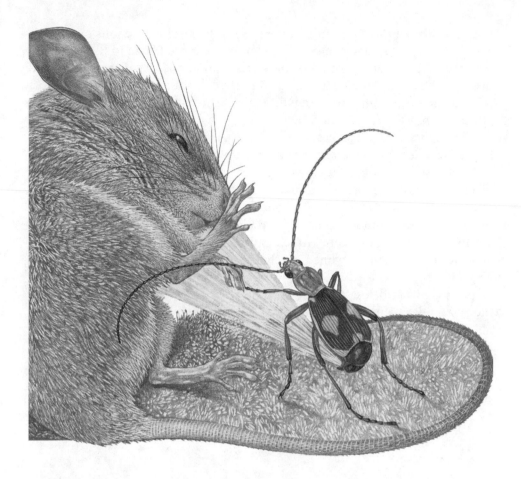

Squirt-Gun Tactics Bombardier beetles are capable of using a highly irritating, hot, repellent spray, aimed accurately by a nozzle located at the tip of its abdomen. Even the predatory grasshopper mouse—which eats scorpions—can't beat the defenses of bombardier beetles.

ated. In fact, there is enough heat to boil the mix. At the same time, the phenols are oxidized to form highly irritating quinones. The pressure builds in the thick-walled reactor and generates enough force to squirt the spray several inches. Even carnivorous, desert-dwelling grasshopper mice that feed on scorpions are repulsed by the bombardier's spray.

The Spiders Come Ashore

At the same time that the ancestral scorpions landed, some now-extinct ancestors to the modern spiders lumbered ashore, armed with a survival blueprint that has allowed them not only to survive a number of extinctions but also to eventually adapt to almost all environs from the Arctic Circle to the fringes of Antarctica. They can be found below sea level, on bone-dry desert floors, and over 20,000 feet up in the Himalayas. An incredible number of these predators inhabit the planet, preying on anything they can subdue. The objects of their attack are mostly insects and other spiders, but they sometimes even prey on small vertebrates. The smallest of the modern spiders is pinhead sized, and the largest, with legs akimbo, can cover a dinner plate.

Their success is attributed to a plastic, or flexible, genetic program, that produces a vast number on variations of the basic plan and to some unique weapon systems, chief among which is the ability to produce and spin a form of silk, the strongest known fiber. (The spinning skills of spiders are responsible for their biological name, Arachnoida. According to Greek mythology, an egotistical young weaver named Arachne, confident in her skills, challenged the goddess of wisdom, Athena, to a weaving contest, which Athena won easily. Depressed by her defeat, Arachne hung herself, but Athena brought her back to life as a spider.) Part of their success is also attributable to their eight remarkable legs which have a total of 56 joints. Spiders can run; climb; jump; and move forward, backward, and sideways and can even use the clawed tips of their legs to run on the dry thread components of their webs. A hooked tip at the end of the leg is used to guide the silk into proper position during web construction.

Webs

The silk glands, located in the spider's abdomen, continuously synthesize liquid proteins, which turn into threads when exposed to air. The ducts from these glands converge in abdominal clusters of nozzles, called spinnerets, from which silk is pulled by the spider's rear legs. There are several types of silk glands, each manufacturing its own type of silk, which may then be used alone or with other types of silk to produce a variety of threads. The silk is used to produce threads large and small, horizontal and vertical, sheetlike or lacelike. It is used to line nests; wrap up eggs in sacs; produce drag lines; or swathe, bind, and immobilize prey. Some spiders, such as the ambush-hunting, ogre-faced spider, can throw nets of sticky threads accurately enough to snare any insect unfortunate enough to come within range. Other spiders dangle lures in the form of blobs of webbing impregnated with sex-attractant species-specific pheromones to bring amorous male moths to their doom.

The silk spiders produce is incredibly strong, estimated to be 20 percent stronger than steel wire of the same diameter (which is a silly comparison because we can't make steel wire that thin). Many web-building spiders construct traps (or webs) of extraordinary geometric beauty and precision, using both dry fibers and sticky fibers. The web design is specific for each species of spider; its construction program is directed by its genes and then hardwired into the spider's brain. Spiders reared in isolation will, when mature, create precisely the same web shape as other members of their species. The size of the web and the distance between strands is dependent on the length of the spider's legs, but orb weavers, large and small, always preserve the structure characteristic of their species. Experimentally, it has been shown that some of the same psychoactive drugs that affect human brains can alter the neural processing responsible for spider's web architecture. In other words, doped-up spiders produce bizarre webs.

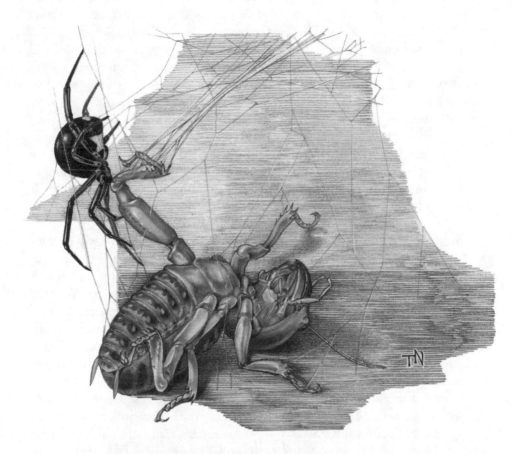

Basic Black and Widowed   Black widow spiders, of the genus *Latrodectus*, are especially efficient predators whose fangs deliver a highly potent nerve-muscle poison which is capable of killing creatures much larger than themselves. There are even rare (but well documented) cases of human deaths.

Some spiders make three-dimensional webs, and others, messy tangles. Some species even cooperate socially, producing mazes of overlapping webs. The webbing is not only strong and sticky but also stretchy and has proven to be diabolically effective as a trap. Most spiders wait patiently until a prey blunders into the web and then, detecting the vibrations with sensors on their feet, rush out and wrap up their prey in silk after immobilizing it with a neurotoxic venom injected with their sharp, curved fangs. Their poisons are a mix of proteins, each having its own effect.

In temperate climates worldwide, in dark, moist places, the shiny black members of the genus *Latrodectus* can be found attending their sparse, messy webs. The most notorious of these is *L. mactans,* the black widow, whose name in Latin translates to "murderous biting robber." Only the female bites, and it can be recognized by a characteristic orange-red, hourglass-shaped design on its ventral abdomen. It is called "widow" because sometimes it eats its smaller mate. This behavior, however, is not exclusive to black widows. Many species of spider will not only eat males of the same species but will cannibalize sisters, mothers, and even their own offspring. There have been several cases in which black widow bites have caused death in a human, but usually the bite produces only an array of painful neuromuscular symptoms. Most bites occur on the hands or arms when a person is out gathering wood, but a significant number of bites occur in outhouses. Black widows tend to build their nests on the underside of the hole in the seat, where they can capture flies attracted to the droppings at the bottom of the pit. Unfortunately, males, when seated in the outhouse, have parts that dangle below the edge of the hole, disturbing the black widow's repose. Being thus provoked, the spider envenomates a very tender body part.

Another group of dangerous spiders that have now spread widely in temperate climates are the shy, brown "hit-and-run specialists" of the genus *Loxosaeles,* more commonly called brown recluse spiders or violin spiders because of the dark violin-shaped pattern on their backs. These medium-sized, six-eyed spiders thrive in dark places and usually attempt to flee when they detect vibrations. They bite only when inadvertently trapped in clothing. Their venom, different from the black widow's, is a mix of tissue-destructive compounds that cause blistering and ulceration of the wound and, sometimes, blood cell destruction. The wounds don't heal well, but are seldom lethal.

The most fearsome looking of the spiders that can cause human envenomation are the wolf and bird spiders, commonly called tarantulas. The largest bird spiders are found in Central and South America and are equipped with 3/4-inch-long fangs, large poison glands, and hairy bodies. The hairs on the abdomen, called urticating hairs, are sharp, barbed, and coated with a potent irritant. Most importantly, they can be shed. On penetrating tender mucous membranes, these hairs are potent deterrents to a would-be attacker. Tarantulas live in burrows, often abandoned rodent tunnels, and lie in wait for prey to trigger the lines of webbing they have laid on the ground. Usually nocturnal hunters, they have a light-reflecting layer on their eyes to enhance their already excellent night vision. When you shine a flashlight into one of their lairs, the eyes glow brightly. Ambush-style hunters, they stalk mainly insects; however, their venom is potent enough to do in small vertebrates. Sometimes small burrow-hunting poisonous snakes, such as the fer-de-lance enter a tunnel looking for mice and end up being dispatched and eaten by a giant spider. The tarantula's venom contains powerful protein-digesting enzymes that turn the snake's tissues to an edible goo which the spider sucks in, leaving only an empty skin.

These big spiders are not particularly aggressive. In recent years, they have even been collected and sold in pet shops. Although large and well armed, tarantulas frequently fall victim to a solitary, large, steely-blue, ground-hunting wasp called, appropriately, tarantula hawk. In most confrontations, the wasp wins, and the paralyzed spider, still alive, is pulled into its own burrow where the wasp lays a single egg on its victim and then departs. The spider has then become the food supply for the developing wasp larva. Male tarantulas are also often killed by females. Mating, then, becomes a life and death adventure.

A number of spiders are active visual hunters, capable of seeing prey with acuity at distances of 10 to 12 inches, a long range for a small animal. The wolf spiders are fast and agile pursuers, while the jumpers either lie in wait or stalk prey. They are capable of prodigious leaps, executed so rapidly that they can barely be followed by the human eye. Millions of years of evolution also produced a variety of deceptive tactics. Some spiders look like twigs, some are colorful and advertise their toxicity, and others blend with their backgrounds. Over the millennia, a number of species of spiders evolved both structural and behavioral adaptations that mimicked ants, whose defenses deter would-be predators. They not only developed a pseudo–wasp waist and patterned their movements to copy those of ants, but one species actually looks like an ant carrying a dead ant.

Often web-building spiders leave their paralyzed, wrapped-up prey and go off hunting for more insects to stock their larder. Some small, stealthy species of spiders exploit this waiting food supply and can walk on the host's web without causing vibrations. There are even some species of spiders that use vibration to lure residents of the web to come to them. These deceptive creatures vibrate the web as if an insect had become entangled in it. When the host arrives, it becomes the hunted.

Chapter 4

The Vertebrate Landing

In the Beginning

Vertebrates, those animals with backbones that followed the invertebrates ashore, were faced with many of the same problems that their invertebrate predecessors had. They needed modified breathing equipment, limbs, and kidneys. Some prototypes were already available to them. The lungfish, as their name implies, had already developed organs with which they could obtain atmospheric oxygen. This profound and momentous breakthrough did not occur suddenly but was produced over millennia by a protracted series of small design changes that made them able to exist on land. Another necessary adaptation to the terrestrial habitat was the ability to walk on land; that equipment was in existence prior to terrestrial life in a group of sluggish lobe-finned fish, or coelacanths, a few of which persist to this day unchanged by the passage of hundreds of millions of years. The coelacanth's four fleshy, lobelike fins had a central axis that we now believe was the basis for four-legged land creatures, or tetrapods, as all land-living vertebrates are called.

The true ancestral tetrapods were the amphibians. During the Devonian period, they established their beachhead on land; but they retained their need for water and were only partially adapted to life on land. Although they had lungs, they still also derived oxygen from their moist skin. They never completely left water behind: they were prone to dehydration and they still needed to lay their eggs in water. However, they did introduce the basic limb plan needed for fully terrestrial life. Although the early amphibian limbs were not particularly well designed for locomotion on land, they all had the same basic bony structures attached at the shoulder and pelvis: a heavy single bone jointed to a pair of thin bones, which articulated with a collection of small bones (the wrists or ankles) and terminated in five jointed digits (fingers and toes) that had terminal claws or nails. The design plan was sufficiently plastic that by discarding a few digits here and expanding a few digits there, the tetrapod limb could be adapted to make a variety of specialized feet—paddle-like structures, such as the flippers of dolphins, and wings.

Another problem for vertebrates on land was sound and vibrational detection. Vibrations in water led to the development of a multitude of sensors, but vibrations in air are too feeble to be detected by the type of apparatus used by fish. The amphibians overcame this problem by developing large, membranous structures that could dramatically amplify the small changes produced when sound waves hit the eardrum.

The Devonian period, 480 to 360 million years ago, when the early ancestral amphibians invaded land, was characterized by seasonal droughts and drying out of shallow lakes and ponds. The drying out produced selection pressures favoring the survival of animals that were efficient walkers, and the fossil record shows that a considerable variety of amphibia existed. For a brief period of Earth's history, almost 100 million years, they were the top predators in their swampy habitats. Some, such as the crocodile-like *Eryops,* strolled ashore with a clumsy, sprawled gait. But most of the amphibians, chained to water for reproduction and prone to desiccation during droughts, became extinct about 245 million years ago, when some cataclysmic event wiped out 90 percent of all living species. Today, the only surviving amphibians—the frogs, toads, and salamanders—are still dependent on freshwater for their survival and have developed some remarkable defense systems, from the toxins in their moist, glandular skin to some exceptional protrusible, guided-missile-like tongues to the capacity to make spectacular leaps to escape predation or to capture prey.

Amphibians—Masters of Chemical Defenses

There are about 2,600 species of modern amphibians: they include frogs, toads, salamanders, and newts. Many of them must be considered among the most poisonous of animals. Their toxins, produced continuously by skin glands, are purely defensive in nature. Their moist surface skin was an ideal place for the growth of bacteria and infective fungi; without question, the toxins evolved as antibacterial and antifungal defenses. Their importance as antimicrobials was demonstrated in laboratory experiments which showed that very low concentrations of frog skin secretions inhibited microbial growth. Further proof was provided by studies on frogs that had been detoxified: they all died of skin infections in a matter of days. Chemical analysis of amphibian toxins reveals a potpourri of over 300 molecular types. They exist in many different genera of amphibians, we will consider just one group of frogs, one group of toads, and one group of salamanders that are so poisonous that they have been known to kill humans.

A Hunter's Chemical Defense System The marine toad is a plump, slow, voracious predator, whose glandular skin secretes a potent witches' brew of toxins. Many amphibians (toads, frogs, and salamanders) possess a remarkable array of defensive chemicals with which to deter potential predators.

Toads (Bufo)

Contrary to popular myth, touching the bumpy skin of a toad does not cause warts. Their secretions, however, can produce serious poisoning. The most toxic of the toads are the large marine toads *(Bufo marinus)* and the Colorado River toads *(Bufo alvarius)* which are responsible for the deaths of many pet dogs who mouthed them and several humans, mostly children.

Their witches' brew of poisons contains an amazing variety of chemicals, including a potent hallucinogenic compound, which is also produced by some of the "magic" mushrooms used to alter brain function. Among these chemicals too is a potent heart poison whose action is similar to that of digitalis derived from the foxglove plant.

Poison Dart Frogs

Among the most lurid and lethal of all animals are the small (1 to 3 inches), colorful dendrobatid frogs of the Central and South American rain forests. Brilliantly colored in reds, yellows, and blues, sometimes with black spots and stripes, sometimes in uniform bold color, they announce their toxicity to would-be predators. Even the ravenous, poisonous giant tarantulas that feed on other frogs will leave the dendrobatids alone. The bright yellow frog *Phyllobates terriblis,* which is 1 to 2 inches long, contains enough skin poison to kill thirty adult men and can even be lethal to the touch. Its huge armory of deadly poisons was recognized by the Choco and Cuna Indians who toxify their tiny blow gun darts by wiping the tips on the back of a captured *P. terriblis.* The poison is so fast acting that its prey, arboreal monkeys, birds, and other small mammals, die almost instantly. Among the multitude of toxins produced by the frog's skin are a number of alkaloids that increase the permeability of cell membranes of nerve and muscle, causing irreversible paralysis. Other of their toxins are stimulants similar to cocaine or sedatives 400 times more potent than morphine. Some others are potent anesthetics.

Of the 135 species of dendrobatids only 55 are toxic, but when removed from their native habitat, even toxic frogs experience depletion of their chemical arsenal. Today, there are pet lovers who specialize in exotic creatures such as the poison dart frogs and their captive laboratory-bred offspring, which soon become nontoxic, possibly due to the fact that their diet in a terrarium does not contain the kind of insects they would normally eat in the wild.

Salamander Poisons

Since ancient times, people have known that salamanders could be poisonous, and tales of their toxicity have been incorporated into their mythology. Recent studies on newts' and salamanders' skin secretions reveal that the poisons they produce are steroids. The basic structure of the steroids is similar to cholesterol and the sex hormones, but the potent compound steroids are nerve poisons that cause convulsions and paralysis.

Another group of salamanders of the genus *Taricha* secrete a poison that is identical to tetrodotoxin (TTX), the poison found in the deadly stonefish, puffer fish, and blue-ringed octopus. How evolution produced the same toxin in such diverse phyla is unknown, but we do know that TTX is a highly potent blocking agent for sodium channels in the membranes of nerve and muscle. There have been confirmed human deaths due to ingestion of the California red-bellied newt *(Taricha torosa)*. In both cases the salamander swallowers were drunken fishermen showing off for their companions.

As predators, most amphibians are ambush hunters. Their targeting system is their remarkable eyes. Once their eyes have locked onto the target, the visual input is processed in the circuits of their retinas and then relayed to the brain. There the information is analyzed for distance and direction of movement and relayed to motor command centers, which dictate appropriate attack signals. In the case of large targets, the salamander gets its prey by a quick charge or jump and a bite. Some very large frogs and toads can even capture and subdue snakes. They can launch themselves at a prey by means of their long, powerful rear legs, which also allow them to make prodigious leaps to escape from predators. In addition, frogs, toads, and some salamanders can remain motionless and hit prey at a distance with their protrusible sticky tongues that are visually aimed.

The Arrival of the Reptiles

Reptiles (meaning crawlers) evolved from some ancestral amphibians about 340 million years ago. These small, lizard-like pioneers of land were the first vertebrates equipped to fully exploit terrestrial life. Their tough, scaly skin could resist drying out, and their eggs had hard mineralized shells that resisted desiccation. In the subsequent diversification of reptilian types, the ancestors of dinosaurs, birds, and even mammals had their humble beginnings. About 248 million years ago, some sort of cataclysmic event caused mass extinctions of about 70 percent of the species. Afterward, the reptilian survivors quickly diversified, multiplied, and took over the planet in the Mesozoic era. The Mesozoic era, which lasted some 155 million years, is also called the Age of Dinosaurs. It was divided into three periods, the Triassic, the Jurassic (popularized by Steven Spielberg), and the Cretaceous.

Many of the new reptiles stood upright and walked and ran in a manner unlike the sprawling gait of their crocodile-like ancestors. This new improved gait, with knees tucked in directly below the body, required the evolution of major changes in the bones of the hips, shoulders, forelimbs, hind limbs, and feet, as well as the development of hinges like ankles and wrists. With these changes dawned the Age of the Dinosaurs, the Triassic period (248 to 213 million years ago). The landmass at that time was a giant supercontinent, now referred to as Pangea, but great plates of the earth's crust were on the move and the landmass was already beginning to split in two. The climate was warm and moist along the coasts, where moisture-loving ferns, horsetails, and treelike cycads provided sustenance for herbivores, while the inner, rainless portions were vast deserts. Most of the reptiles of the Triassic period were small, although there were a few larger plant eaters. There were small mammal-like reptiles that gave rise to the tiny shrewlike prototherians, the ancestral mammals. There were also ancestors of the birds and ruling dinosaurs that evolved in the Jurassic period (213 to 144 million years ago).

During the Jurassic period, the continents were drifting apart. Old mountains were wearing down, and shallow seas invaded much of the landmass, bringing rain to former deserts. The warm climates promoted diversification and growth, including the appearance of a number of enormous herbivores and predators. By the beginning of the Cretaceous period (*Creta* is Latin for "chalk") the continents had almost reached their present positions, although what is India today was still an island on the move. The climates had become seasonal, and an enormous variety of flowers, plants, and trees covered the land's surface. The Cretaceous period (144 to 165 million years ago), the longest part of the Age of the Dinosaurs, came to an end at the Cretaceous-Tertiary boundary when a massive extraterrestrial body impacted the earth in the region of the Yucatan and wiped out the vast majority of species. But during their protracted reign, the dinosaurs had already established the strategies and tactics for the arms race to follow.

Reptilian Predatory Strategies and Weaponry

The reptiles' methods of attacking prey varied with their size, speed, agility, and potential victim. All were in part reliant on their biting mouths, most of which were equipped with sharp teeth. Some, like the tyranosaurids, had immense, tooth-lined jaws that could open to an enormous gape. Their skulls had

Fleet-Footed, Cooperative Hunters (facing page)   The massive *Iguanadon* is about to succumb to persistent attacks by a pack of *Velociraptors*, whose primary weapon is a long, razor-sharp claw on one of their hind toes. Slashed repeatedly by this lethal weapon, the *Iguanadon* will eventually bleed to death.

evolved a number of ridges for the attachment of jaw-closing muscles. Some of these skulls were over 6 feet long and had jaws equipped with ten-inch-long steak-knife-like teeth that could rip great chunks of flesh from their victims. Other predators had elongated tooth-lined snouts that allowed them to make a sudden snatch at fish in shallow waters. Some, such as *Oviraptor* (which means "egg thief"), had short, toothless jaws with specialized sharp bones pointing downward from the roof of its mouth; these structures could punch through exceptionally tough egg shells.

While the larger meat eaters had clawed forelimbs that were too short to do much damage, others such as *Deinonychus* and *Trodon* had very very long arms with three talon-tipped fingers for grasping, tearing, and slashing prey. Although most of these upright predators had claws on the toes of their hind limbs, it is doubtful that they used them as weapons. But several varieties of *Deinonychus,* such as the 6-foot-long *Velociraptor* (the "quick plunderer"), had a particularly formidable weapon on the second toe of each of its rear feet, a retractable, 4-inch-long saber-like claw. These bipedal fast-running hunters were keen eyed, and probably had three-dimensional color vision as well. Their relatively large brains, with well-developed forebrains, enabled them to learn and also provided them with hand-eye and foot-eye coordination. They were cursorial (chasers), so they could outrun slower much larger herbivores that may have outweighed them more than a thousand times. More than likely, they hunted in packs, using coordinated tactics similar to those of wolves and hyenas of today. They also may have had considerable stamina and persistence but were probably too small to be able to give a killing bite. In reconstructing a hunt scenario, we can speculate that they probably leaped, hung on, and repeatedly slashed with the second toe claw until the prey bled to death. Clearly, they must have been truly awesome predators. Recently some new fossil finds in Argentina recovered a foot-long second toe claw that belonged to a *Velociraptor*-like creature six times larger than its cousins. These same digs also revealed fossils of a tyranosauid larger than any previously known—*Tyrannosaurus rex.*

Tyrannosaurus rex, probably our archetypical predatory dinosaur, was present in the later third of the Cretaceous period. Some may have measured 40 feet in length from head to tail and weighed as much as 7 tons. They were not particularly fast; they didn't have to be, because their prey were usually even larger and slower. It has been suggested that they hunted cooperatively and usually went for the head and neck, as modern big cats do. Given the size of their rear legs, they were capable of making a high-speed charge, but given their bulk, they probably could not pursue at speed. While the tyranosaurids' brains were proportionately smaller and less complex than those of dinosaurs like *Velociraptor,* they were infinitely more advanced than those of most of their pea-brained prey.

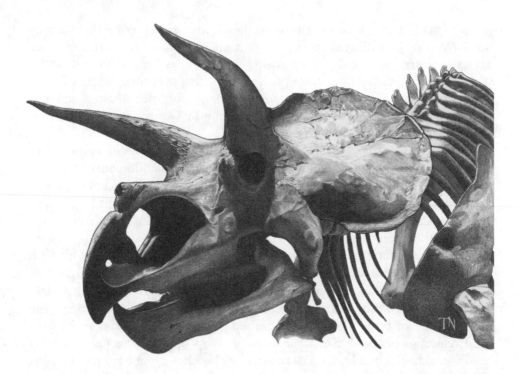

Horns and Armor   Many herbivores evolved extensive armor to protect themselves from predators. The extinct *Triceratops* not only boasted a massive frontal armored shield, but added to its defenses with long, pointed horns. Most predators facing the charge of *Triceratops* would back off and look for easier prey.

Reptilian Defense Strategies and Weaponry

One tried and proven evolutionary strategy for dealing with predators is to grow big enough to deter them by intimidation. The largest of all land dwellers were the terrestrial sauropods of the Jurassic and Cretaceous periods. Some weighed almost 100 tons, had 40-foot-long necks and 30-foot-long tails. These slow, lumbering giants traveled in herds and could use their powerfully muscled tails as flails to deter small predators. Their only other defensive weapons were immense pointed toe and thumb claws. A backward kick from a sauropod could seriously wound a predator attacking from the rear, and a head-on attack might have been met by the sauropod's rearing up and brandishing its thumb claws to stab at the predator. However, these great, plodding herbivores may not have been coordinated enough to lash out effectively at faster and more agile predators, because they had smaller, less complex brains than any other vertebrate. Furthermore, selection processes working over time produced huge predators to prey on them. Great size as a basis for specialization has not proven to be a successful long-term strategy, and the sauropods' sheer bulk was not enough to ensure their survival.

Another way of coping with flesh-eating predators was the development of body armor imbedded in tough skins. Many herbivorous dinosaurs had bony studs, bony plates, and a great variety of bony spines and sharp spikes to protect themselves. Many predators attack their fleeing prey from behind and some herbivores, such as the stegosaurids, evolved sharp, long, bony tail spikes with which to defend their rears. Other armored prey, such as the ankylosaurids, developed large, bony clubs at the tips of their powerfully muscled tails. A direct hit with such a weapon could easily wound or kill a predator. Some of the later large plant eaters had huge heads, rhinoceros-type bodies, and an enormous, bony protective frill equipped with horns protecting their head, neck, and shoulders. The most familiar of these was the three-horned *Triceratops* which appeared in the late Cretaceous period. It had a pair of long, sharp-tipped horns near the forehead and a short, blunt horn on its snout. In all likelihood, these animals traveled in herds, grazed on vegetation, and migrated. It also seems very likely that they used their defensive weaponry to charge at threatening predators and were effective. But their success against predators couldn't protect them from the asteroid impact that caused the great extinction.

Reptilian Survivors of the Great Extinction

Today's crocodilians, turtles, tortoises, lizards, and snakes evolved from ancestral forms that survived, diversifying over the past 60 million years. Although the defensive and offensive tactics, strategies, and weaponry they developed to meet the needs of their ever-changing environments varied widely, almost all were cold blooded and were egg layers. The success of body armor such as that seen in the large land tortoises found on the Galapagos Islands is attested to by the tortoises' longevity of over 100 years. Modern reptiles evolved strategies and tactics similar to those of other animals. One tactic, the use of which is widespread, is deception. Some snakes' tail tips, modified to look like insects, are dangled in front of their mouth to lure prey into range. Some species of large snapping turtles have tongues with wormlike, wiggling appendages which lure unsuspecting prey into their gaping jaws.

On the defensive front, reptiles have developed a number of successful ploys. Some lizards fleeing from a predator can, as we have noted, voluntarily drop a part of their tail, which continues to wiggle and distract the predator. Many snakes, when threatened, feign death. There is even a lizard that squirts blood out of its eyes in an attempt to distract or deter predators. Since most snakes and lizards are fragile, vulnerable creatures, natural selection has provided them with a veritable arsenal of defensive equipment. Some poisonous snakes advertise their toxicity with black, yellow, red, and white banding patterns, which have been mimicked by some nonpoisonous species. An auditory warning meaning, "Don't tread on me" is sent by rattles at the end of the tails of the crotalid snakes; other snakes hiss loudly, and the coral snakes, when threatened, turn out their anal sacs with an audible pop. Cobras rear up and spread their hoods marked with a pigmented design that looks like bold spectacled eyes, and many cobra species can spit a spray of toxic venom, aiming at the eyes of an intruder. This deterrent to aggression can be effective at a range of 10 to 12 feet. Many lizards and snakes also have repugnatorial glands on either side of the anus; anyone who has collected snakes and lizards can attest to the vileness of the stench of their defensive defecations.

Snakes

Dangerous reptiles have long been deified by humans in the hope that elevating the creatures to godlike status will protect humans from harm. A number of ancient cultures revered crocodiles. The Egyptians built temples to the half-human, half-crocodile god, Sobek. They constructed a city they called Crocodilopolis and even mummified thousands of Nile crocodiles. The Egyptians also deified cobras, as did followers of some religions in India. Even today, most people have a revulsion for snakes, particularly for those that are venomous. Of the 2,000 known species of snakes, however, there are only 400 species that are truly venomous. Venomous snakes are widely distributed in all tropical and subtropical zones and in most temperate zones and are responsible for approximately 100,000 deaths per year, mostly on the Indian subcontinent.

The most dangerous of the envenoming snakes belong to the family Elapidae, which includes the coral snakes, cobras, kraits, and mambas. One's chance of surviving a snake's bite depends on the amount of venom injected and the relative toxicity of the mix of poisons. When they bite defensively, snakes do not always envenomate; when hunting prey, however, they always envenomate. Controlling envenomation in this way makes sense, since the manufacture of venom is metabolically expensive and snakes must be miserly in expending their weaponry. Lethality also depends on the site of the bite and the physical condition of the victim.

A Nasty Habit of Spitting   Cobras subdue their prey with a venomous bite that kills quickly. Like most snakes, they are relatively fragile creatures; when threatened, they rear up, flare their hoods, and hiss loudly to deter their attackers. Many cobras also possess the ability to accurately spray a stream of venom into the faces of their enemies.

Elapid bites are responsible for the most deaths; 25 percent of human bites are fatal. The most deadly of the elapids is the black mamba, a nocturnal hunter that, like most snakes, spends much of its time in a burrow out of harm's way. The long (12 feet or more), slender mamba is among the most aggressive snakes. When disturbed, it will raise its body, spread its narrow hood and open its mouth, showing the black part of its throat as a threat display. Mambas strike with little provocation. Exceptionally agile and fast, they can cover short distances at speeds of up to 20 miles per hour. Often they will aggressively move toward an intruder. Their bites are almost always fatal. Knowing this, one African wildlife photographer, when bitten on a finger, rapidly applied a tourniquet and cut off his own finger. This decision undoubtedly saved his life.

Most elapid venoms are a mixture of enzymes and polypeptide nerve and heart poisons that are produced in modified salivary glands located between the eye and the rear of the upper jaw. The venom is carried to pointed hollow fangs on the upper jaw at the front of the mouth. The fangs act like hypodermic needles; the poison is injected through an opening on the side of the fang.

While the mamba is undeniably the most dangerous, other elapids, such as the cobras, are more numerous and widespread and therefore cause more deaths. The large elapids found in Australia are also among the most dangerous venomous snakes while the small, boldly colored American elapids, the coral snakes, are shy and hardly ever lethal.

Of all the elapids, the hamadryad, or king cobra, of Asia is the largest venomous snake known, reaching a maximum length of 20 feet. This enormous snake feeds primarily on other snakes, including a variety of cobras.

Two other families of terrestrial poisonous snakes are widespread. These are the vipers and the pit vipers; the latter group includes the rattlesnakes. The pit viper, or crotalid, is a nocturnal hunter of warm-blooded prey, capable of accurate strikes in total darkness. This ability is due to a pair of remarkable long-range, ultrasensitive heat detectors located on either side of its head below and behind its nostrils and in front of its eyes. These infrared or heat sensors can detect temperature differences as little as 0.5°C at a distance of 10 to 12 inches. Temperature differences between the right and left pits are relayed to the snake's targeting center in its brain, allowing it to line up the target within 5 degrees of dead center. Prior to the strike, the snake holds its neck and upper body in an S-shaped curve, like a compressed spring, and the strike itself is so fast that it can barely be followed by the human eye. As the strike is launched, the viper opens its mouth almost 180 degrees and erects the retractable fangs at the front of its upper jaw. The strike, with the fangs pointed forward in the mouth is like a rapier thrust, usually producing two puncture wounds. Some of the bigger pit vipers, like the diamondback rattlesnake of the southeastern United States, the cascabel of South America, and the very large bushmaster, have enormous fangs that penetrate deeply into muscle, ensuring rapid absorption of the venom and quick knockdown of the prey. The pit viper's venom is a deadly mix of toxins, some of which break down tissue, destroy red blood cells, cause bleeding, and even inactivate a vital antibacterial component of the body's defenses, the so-called complement system. There are also varying amounts of potent nerve toxins, which cause paralysis. The most potent venom is found in the South American rattlesnake called the cascabel, which is a very aggressive and heavy-bodied snake. One herpetologist described it as a "sinister and insolent snake." This snake almost always envenomates and given its long fangs and large size, is probably the most dangerous pit viper.

After envenomating its prey, this snake quickly returns to its coiled position and then begins to follow the scent trail left by its mortally wounded victim. To do this, it repeatedly flicks out its odorant-gathering forked tongue. As the tongue returns to its mouth, scent detectors called Jacobson's organs analyze the chemicals on its surface. Pursuit is leisurely; the prey doesn't get very far before it collapses. The snake then unhinges its jaws and even

A Heat-Seeking Guided Missile   The so-called pit vipers, such as this diamondback rattlesnake, can locate their warm-blooded prey in total darkness by using a pair of exquisitely sensitive heat detectors in the front of their heads. These snakes' lightning-fast strikes are usually accurate within 5 degrees of center. It is not surprising that the U.S. Air Force has named its heat-seeking guided missile after the sidewinder rattlesnake.

unhinges the connection between the two halves of the lower jaw and slowly swallows its prey head first. Since snakes' ribs do not attach to a breast bone, they can swallow prey whose girth is much larger than their own. They may bulge for a while until the meal is digested, and during digestion they usually retire to a safe burrow.

Not all snakes hunt warm-blooded prey. Some poisonous and nonpoisonous snakes specialize in hunting other snakes which, being long and tubular, fit readily into the bowel of a tubular hunter.

Constrictors—The Hug of Death

The largest, heaviest snakes are nonvenomous constrictors. These predators are usually nocturnal hunters and are armed with an array of infrared sensory pits at the margins of their upper jaws. Although their prey are usually warm-blooded mammals, they will kill cold-blooded, medium-sized crocodiles and caymans. All constrictors can stealthily stalk prey. More often than not, though, they are ambush hunters, waiting for prey to come within

A Reptile that Will Take Your Breath Away   The world's largest snakes—pythons, boas, and anacondas—kill their prey with crushing muscular contractions of their bodies. Two or three coils around a prey's chest will literally squeeze the breath from that prey. The python here, shown incubating her eggs, has an array of 13 heat-sensitive pits on its upper lip, which are used to locate warm-blooded prey.

range. Their strike is very swift and its impact alone may be sufficient to stun a victim, but their usual tactic is to bite down on the prey, securing it with their sharp, pointed, backward-facing teeth of uniform size set in both their upper and lower jaws. The constrictor then throws two or three coils of its powerfully muscled body around the victim and tightens the coils until the prey suffocates. The pressure in the prey's chest cavity becomes so great that the veins carrying the blood to the heart collapse. The constrictor then swallows the prey whole. This process can take considerable time; however, a valve on its breathing tube permits the snake to swallow and breathe at the same time.

There are many myths about the size of the giant constrictors. Recently, I acted as consultant for a movie called *Anaconda*, in which the villain was a 40-foot-long anaconda snake that attacked a group of moviemakers in the Amazon. Although such large snakes may have existed at one time, today a 20-footer is considered top of the line. For years, the New York Zoological Society has had a standing offer of a cash prize for a 30-footer, and no one has yet claimed the money. The anaconda is a member of the family Boidae, of which the most well known are the boa constrictors. Their size has also been wildly exaggerated; and the largest reliable record of length is about 18 feet, although a 15-footer is considered very large. Stealthy, slow stalkers, boa constrictors are capable of eating some medium-sized animals such as a 40-pound ocelot.

Digestion is a slow process, and almost everything is digested, including bones. As they are endowed with a slow metabolism and are relatively inactive, constrictors can go for long intervals between meals.

The pythons are the Old World equivalent of the New World boas. The largest, the reticulated python, may reach over 30 feet in length but is less heavily bodied than the anaconda. Some, such as the Indian rock python, almost always hunt at the water's edge, and all pythons are good swimmers. Some species are arboreal and others are burrowers, hunting underground rodents. They may capture and eat several small prey in succession. Pythons are also devoted mothers, guarding their clutch of up to 100 eggs and wrapping themselves around them to provide the heat needed for egg development during the 2- to 3-month incubation period.

The Lizards

There are many different patterns of living among the 3,000 kinds of lizards. Some can run on water, one species can squirt blood out of its eyes, and many can change their color to match their background. We examine three types in detail: the chameleon, the poisonous beaded lizards, and the crocodile-sized terrestrial dragon lizards.

About 80 species of chameleons exist, all endowed with a number of remarkable features. Most have long, prehensile tails that they can wrap around a twig, augmenting the tenacious grasp provided by their unusual toes. These toes, three on the inside of the front foot and three on the outside of the hind foot, act like tongs. When chameleons move, which is seldom, they creep forward slowly by moving one front foot and one opposing rear foot simultaneously. Their ability to change color to match their background is legendary, and their slow stealthy movements and adjustable camouflage are ideal for ambush hunting. They spot their targets, usually insects, with turreted, scale-covered eyes that swivel independently. As both eyes look forward, they relay a stereoscopic image to the missile-aiming fire-control center in the chameleon's brain. To improve range finding, the chameleon rocks from side to side to view its intended prey from several angles.

Once locked onto a target, the chameleon sends commands to its guided missile system, a protusible, sticky tongue whose range is longer than the chameleon's body. Initially, the tongue protrudes in preparation for launching. Then, with remarkable speed, it shoots out, propelled in part by the contraction of muscles attached to a jointed, V-shaped bone and of circular muscles at the thickened tongue tip. These fast-twitch muscle fibers allow for tremendous acceleration, so quick that it can be followed only by means of high-speed cinematography. The adhesive tongue tip secures its target, which is rapidly

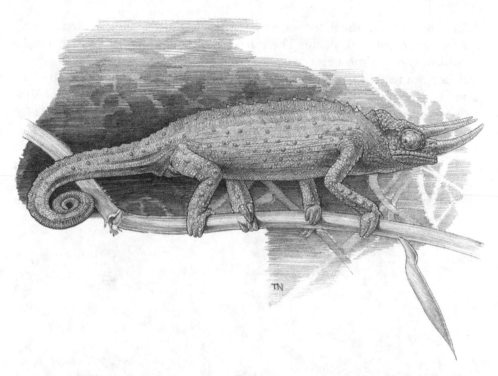

An Optically Guided Missile The chameleon has turreted eyes that are capable of moving independently while exploring their visual surroundings. But when these eyes detect a prey, the lizard fixes its gaze in such a way that it gets a binocular image of the target. It then rapidly shoots out its long, sticky tongue, which hits the target (often a spider or an insect) with amazing accuracy.

returned to the mouth by a different set of muscles. Most chameleons are only a few inches long and specialize in eating spiders and insects, but some of the larger chameleons, up to 20 inches long, can catch small birds, mammals, or other lizards.

Only two species of lizards are venomous, the massive-headed, short-legged, belly-dragging Mexican beaded lizard, *Heloderma horridum,* and the Gila monster, *Heloderma suspectum.* The latter was so named because its venomous nature was initially only suspected and not proven. These two slow-moving, colorful (with yellow, pink, and black beaded scales) creatures use their flicking tongues to locate prey. Odorants picked up by the tongue are carried to a taste-smell detector, Jacobson's organ, located in the roof of the mouth. Their favorite foods are eggs and young nesting rodents. Although sluggish, they can rear up and hiss loudly when confronted and can bite with a rapid sideways movement. The bite is viselike and very difficult to disengage.

While hanging on, the lizard moves its jaws from side to side and releases its venom, a potent neurotoxin. There have been a number of human deaths recorded, and in almost every case the victim of the bite was drunk. Gila monsters are not recommended as pets, not only because of their potent venom but also because of their incredibly foul smell. I once had to drive from New York to Boston in a snowstorm with a 2-foot-long, flatulent Gila monster in a cage in the car's back seat. With the windows closed, the animal's stench was overwhelming and unforgettable.

The largest of all modern lizards, the Komodo dragons, are only found on a few Indonesian islands and are thought to be descendants of the extinct giant marine mosasaurs that died out 65 million years ago. Large

Dragon Lizards: Living Fossils   Komodo dragons are the largest modern lizards. Fierce carnivores, they can weigh up to several hundred pounds and are capable of rapid acceleration in pursuing prey. They have an exceptional sense of smell, with which they can detect live prey as well as carrion. Interestingly, the Komodo dragon has a third eye (located in a hole on the top of its head) which is covered by a transparent layer of skin. This third eye probably serves to synchronize the lizard's physiology with daily and seasonal biological clocks.

male dragons can weigh up to 300 pounds and measure 10 feet in length. These thick-necked, long-headed, slow-moving lizards are daytime hunters capable of killing deer and pigs. They locate prey by using their constantly flicking forked tongues to pick up the animal's smell. Most of their food is carrion, but as ambush hunters, they can make a sudden charge, grasp prey with their long claws, and deliver a killing bite. Gluttonous eaters, large Komodo dragons don't share their food with smaller ones. Using powerful sideways sweeps of their muscular tails, they bludgeon their smaller competitors.

The Mammals Take Over

Sixty-five million years ago, inconspicuous shrewlike mammals that had evolved from an ancient mammal-like reptile were afforded the opportunity to evolve, diversify, and eventually dominate Earth's surface. These inconspicuous, small, hairy ancestral mammals were endowed with several innovations that laid the groundwork for their eventual evolutionary success. They were warm blooded; bore their young internally within a fluid-filled sac, the placenta; and had the capability of nursing their young on a highly nutritious liquid, milk. Not all mammals had placentas, and some ancient mammals, such as the duck-billed platypus and the spiny anteater of Australia, still laid eggs but did nurse their young.

The mammalian brain was also different from all earlier vertebrate brains in that it was proportionately larger and more complex. Within this brain, there had evolved a collection of centers and connecting communication channels, which dictated instructions for motivated behaviors involved in preservation of both the individual and the species.

This complex neural information-processing system was most advanced in those predatory mammals whose survival depended on the acquisition of an array of hunting skills. The higher an animal was on the evolutionary ladder, the greater was the development of the brain's control systems.

Evolution proceeds by branching and small changes rather than major transformations. With these changes, the basic genetic blueprint is altered in ways that are adaptive for specific environmental situations. Over the course of time, if special tools were needed, they were invented by modifying older, more primitive structures. This is how a complex of circuits arose from the old vertebrate olfactory brain called the limbic system. First described by Paul McLean, the limbic system gave rise to a hierarchy of connected structures that regulated approach-avoidance and fight-or-flight behaviors. Included in these circuits were connections to the memory banks of the thinking brain, where past experiences involved in risk-benefit and cost-benefit decisions are stored.

The limbic system has a collection of anatomically discrete centers regulating feeding, drinking, temperature, fighting, reward and punishment, mating, and care of the young. Some are appetite centers for aggression; natural selection favored the evolution of aggressive and competitive animals. The centers for aggression and fighting behavior were carefully mapped out by the Swiss Nobel Prize winner W. R. Hess, who found that there were also passivity centers which were associated with them and could turn off aggressive behaviors. The centers for aggression do not dictate attack but rather put the animal into a heightened state of alertness and preparation for attack. The attack decision comes from the thinking components of the brain. For example, if electrodes are implanted in the so-called rage centers of a cat and these centers are stimulated, the cat immediately goes into an attack set. Its claws unsheathe, its fangs are bared, its back arches, and the hair on its nape stands on end. Its attentiveness and focus become acute. The blood flow to its brain and muscles increases, as does its muscle tone. Emergency hormones are released in a torrent. It is ready, but it does not attack. However, if it is then presented with an enemy, such as another cat with which it has previously battled, the attack is activated with all-out fury. Having succeeded in driving away or killing its target, the victor returns to a totally passive state, satiated with venting his rage, due to signals from the passivity center turning off the rage center. Similar responses are seen in other appetite-satiety centers, such as those for feeding and sex. Fulfillment of an appetite, then, is rewarded by some chemicals in the brain that make the animal feel good. In this way, natural selection favored the evolution of a mechanism or program that rewarded success and punished failure. Put in human terms, the animal would feel "the thrill of victory and the agony of defeat."

The Call to Battle Stations

Vertebrate brains are constantly receiving massive amounts of information from the sensors that monitor their environment. However, one peculiarity of both sensors and nerve cells that make up the data-processing circuits is that if the inputs are repetitive and monotonous, they become irrelevant. Sensors undergo a process called adaptation, and nerve cells habituate. They can still do their job, but their sensitivity is remarkably diminished. In other words, they get in a rut. Predators stalking prey have evolved a variety of stealth techniques to approach close enough to their prey without activating adapted sensors or habituated neural circuits. Lions (or any stealthy predators) move toward their prey with great patience and precision, lest they activate the prey's alarm system.

The alarm system in vertebrate land animals is an evolutionarily ancient structure in the brain stem. All sensory inputs not only send signals to the sensory part of the brain for analysis but also send signals to the brain stem's reticular activating system, a structure that Guiseppi Morruzzi and Hoarace MacGoun called "the waking brain." Any novel or strong stimulus will dishabituate the brain, shaking it out of its routine. The initial response is literally a call to battle stations. First, a nonspecific arousal is activated and the animal is alerted. Its eyes open wide and start to scan its surroundings; its muscle tone increases and it is behaviorally alert. Now, the animal is able to identify and focus its attention on what is relative to its survival. The reticular formation is selective; that is, it amplifies relevant sensory information and suppresses irrelevant information. The animal now concentrates on the threat. At the same time, its alerted behavior signals to the predator, "I see you."

Watching a trio of lionesses hunt, I was struck by their incredibly slow and stealthy approach to a large eland antelope that was resting in an open grassland about 300 yards away. Almost invisible in the tall dry grass that matched their coat color, the lionesses moved in downwind. When they were 200 yards away the eland's head popped up and its ears pricked up as it visually searched its surroundings. Suddenly it fixed its gaze in the direction of the three huntresses and rose to its feet. It was now alert and focused on the pending threat. All three lionesses froze in their tracks but it was too late. Experienced hunters, they knew their limitations. If they moved forward, the eland would take off, and they knew they could never overcome a 200-yard lead by an animal that could run as fast as they could. Since the lionesses knew the game was up, they abandoned the hunt rather than waste energy in futile pursuit.

The reticular activating system is divided into ascending and descending parts that can selectively alter the responses of the brain and spinal cord. Not only does the system govern nonspecific arousal and attentiveness (or focused attention); it also regulates sleep and dreams.

Sleep, The Great Restorer

All vertebrate animals include in their life program a mandatory period of inactivity which gives the organism time to repair the wear and tear caused by the processes involved in survival. Sleep is an absolute necessity, and during its deepest stage, growth hormone is released. Growth hormone stimulates protein synthesis and repair. Animals deprived of sleep become inefficient and impaired; this is why sleep is called the great restorer.

There are several stages of sleep. As the animal sinks into the sleep state, its heart slows and its blood pressure drops, as does its body temperature. Its

The King of Rest Lions hunt singly or cooperatively, with the majority of their kills made by female members of the pride. Stalking and hunting expend a great deal of their energy, and lions spend three-quarters of their time recovering from their strenuous hunts.

breathing slows and becomes deeper and regular. All these changes promote repair and rejuvenation. But one stage of sleep, called paradoxical sleep or Rapid Eye Movement sleep (REM sleep), is characterized by changes in brain waves, heart rate, blood pressure, and respiration more like that seen in an awake animal. During REM sleep, the animal is dreaming. We are not exactly sure what functions REM sleep has, but we do know that an animal deprived of REM sleep and then allowed to sleep uninterrupted will go into an orgy of REM sleep as if it were catching up on something essential for its well-being. One theory advanced to explain this ubiquitous phenomenon is that the animal is unloading pent-up excitement in those parts of the brain having to do with survival, preservation of the individual (feeding, drinking, hunting, and fighting), and those parts having to do with preservation of the species (reproduction and care of young). Clearly for higher mammals, dreams are essential to survival, except in the Australian monotremes which don't experience REM sleep.

The Hunters

All predators are successful only part of the time. If their hunting efficiency were a hundred percent, they would soon eliminate their food supply. Hunting efficiency is highly variable. One of the most efficient hunters is a mouse-sized animal whose very name conjures up the image of nastiness and aggression, the shrew. Unfortunately, this name as construed in an earlier male-dominated society undeservedly became associated with females, and even today human males have difficulty dealing with aggressiveness of smaller females whose tactics don't include many of the restraints found in male-male confrontations. Shakespeare's Kate in *The Taming of the Shrew* epitomizes the human male's attitude, even though shrewishness can be equally applied to some males whose behavior emulates that of this voracious, tiny, high-strung predator.

Shrews locate their prey by their sense of smell and don't hesitate to attack animals ten times their size. They have the highest rate of metabolism of any mammalian predator; their heart beats up to 500 times per minute, whereas the average human heart beats only 70 times per minute. Their attack is typically an all-out, furious assault. Moreover, shrews are the only mammals that use venom to hunt. The venom produced by their salivary glands is a potent nerve poison similar to that of cobras. Shrews frequently invade rodent burrows in their search for food and sometimes in the process encounter small snakes, such as the common garden snake. It's no contest, and soon the shrew begins to feast on its much larger victim.

For every tool invented by predators, prey have evolved counter measures. To survive their predators, some shrews have evolved a potent chemical defense, repugnatorial glands that produce a foul odor and taste. I once saw my large shepherd pounce on a shrew, mouth it, and immediately let it go. The dog then spent the next 10 minutes rubbing its nose in the snow, trying to get rid of the shrew's secretions.

There are other small predatory mammals that, like the shrew, are ferocious at times and also have evolved potent antipredator chemical defences. The most potent stinkers are the skunks and polecats. Often they are boldly marked in black and white to advertise their defenses. Most mammalian predators are highly reliant on their sense of smell and are endowed with large numbers of olfactory sensors. In the sensory portions of their brains there is a point-to-point representation, so that those portions of the brain processing odor information are proportionately larger. This is true for all mammals. The duck-billed platypus's brain has a great portion dedicated to processing sensory information from its bill, and humans, with their marvelous manual dexterity, have relatively greater portions of their sensory brain involved in analyzing touch and pressure signals from the hands. This being the case, it is not surprising that the highly sensitive noses of predators, when sprayed with the skunk's repugnatorial chemicals, experience excruciating discomfort. Not all animals are deterred by skunks' spray; the great horned owl, for one, dines on skunks frequently. Skunks also haven't yet discovered that their spray doesn't protect them from high-speed automobiles, which leave behind, flattened, furry, black-and-white, still malodorous skunk carcasses.

Tunnel Hunters—
The Mouse Like an Arrow

At the end of World War I, the French decided that the best way to protect their vulnerable borders was to build a vast network of underground fortifications along their frontier with Germany. They spent billions of francs on an elaborate system of bunkers, tunnels, and even an underground railroad. Then, lulled by a false sense of security, they hunkered down. The Maginot Line, as the underground defense system was called, was easily cracked by Nazi forces. The German underground equivalents, the Western Wall and the Siegfried Line, fared no better and were overwhelmed by the Allied onslaught.

In the jungles of the Pacific, Japanese underground defenses proved to be more difficult to penetrate. These underground bastions resisted heavy naval gunfire and air attack and could only be overcome by costly "ferreting out" of the defenders by the infantry. The penultimate masters of jungle-floor defense were the Vietcong, whose molelike mazes of subterranean tunnels cost U.S. forces such high casualties that the United States abandoned the war.

Tunnel Hunter   The small—but bold and swift—ermine has a slender, streamlined body that allows it to pursue rodents (which account for ninety-nine percent of its prey) in their protective burrows. It uses these same burrows to avoid attacks by owls, hawks, and other raptors.

In the animal world, many small, vulnerable animals countered the threat posed by potent predators by adopting tunneling strategies, creating elaborate networks of subterranean runways with multiple bolt-hole exits.

However, for every defense system a new mode of predation evolves. A number of animals adapted to hunting in the narrow confines of the tunnels. One of the most efficient of these predators is the weasel. Bold and confident all out of proportion to their small size, these solitary predators are capable of taking prey much larger than themselves. Some are diminutive (as small as 5 inches long). Seldom do they weigh more than 10 to 12 ounces. They are lithe, streamlined hunters whose genus name *Mustela* in Latin means "mouse like a spear." They have broad skulls, powerful jaw muscles, sharp teeth, and long necks, which enable them to carry their prey without tripping on their short front legs. Ninety-nine percent of their prey are rodents, and they have honed their rat-hunting skills for about 4 million years.

The key to their hunting success is slinkiness, speed, stealth, and slimness. They don't store fat; after all, a plump subterranean predator would be at a disadvantage in the narrow confines of a burrow. They have an enormous metabolic rate and their heart beats 500 times per minute. To maintain this accelerated metabolism, they must refuel often. They eat five to ten meals a day, consuming one-third of their body weight. The main problem the rapacious weasels must cope with is a declining food supply. When rodent populations crash (and why this happens is not known), the weasel populations also decline. Conversely, as rodent populations explode, so does the reproduction of weasels.

The world of the weasel is hardly free of danger: a number of larger predators such as foxes, owls, and hawks hunt them. Alert to such threats, their first response is to escape down the nearest mole hole. They also use defensive camouflage, changing their coat color to white to blend in with the snow in northern climates. Their tails, however, remain black. Such a prominent black tail would certainly seem to be counterproductive, but it is actually a form of deception. Aerial hunters such as hawks, seeing the black tip, hesitate for a few crucial moments as they try to figure out the direction to attack. This brief confusion gives the speedy weasel a chance to bolt. Another defensive tactic is to play dead since many predators' eyes are geared to movement. As a final line of defense, weasels, which are related to skunks, can unleash a potent, foul-smelling chemical deterrent that discourages olfactory hunters such as foxes.

Cooperative Carnivores

The early, small flesh-eating mammals made their living eating invertebrates. It wasn't long, though, before some became capable of preying on their vertebrate relatives, and soon the carnivorous mammals became the successors to the predatory dinosaurs. However, before they could do this, they had to invent some new weaponry by modifying the basic mammalian tooth plan. The killing weapons had to be sufficiently sharp and long to pierce tough hides, tendons, and even bones. Grinding teeth were reduced, and those molars that were retained for chewing were modified. At the front end of the jaws were the wedge-shaped nipping tools; behind them were large, sharply-pointed canines, or dog teeth, for stabbing and slashing. The cheek teeth that remained had pointed cusps and sharp ridges for cutting.

Carnivores needed speed to catch their prey either by pursuit or by short-range bursts, requiring rapid acceleration but not necessarily stamina. Their forelimbs were modified to be supple; claws developed for grasping prey. By the beginning of the Oligocene period, modern carnivores—dogs, cats, bears, and others—were preying on large, small-brained, slow herbivores. Eventually, most of the big clumsy herbivores passed out of existence and were replaced by faster prey that could outrun most carnivores, provided they had sufficient warning. However, since the carnivores still had greater plasticity and learning power in their brains they could adjust their hunting strategies, learning by experience. The process of gaining this experience varies widely. A newborn ungulate such as a wildebeest or white bearded gnu learns its identity and basic survival plan in the few critical hours after being born. On the other hand, cheetah, lion, and tiger cubs have to spend at least two years of training before they become sufficiently experienced to survive on their own.

There are two main groups of terrestrial carnivores: the cats, which include the doglike hyenas, and the dogs, which include weasels, raccoons, bears, wolves, and cape hunting dogs. Both groups developed two specialized hunting strategies. Most are solitary hunters; but lions, hunting dogs, hyenas, and wolves developed the capacity to hunt cooperatively.

The Pride

The largest of the cooperative cats are the lions, whose entire way of life is a rigidly choreographed program of togetherness. The basic unit is the pride, which is composed of one or more related males, a number of females, and their cubs of various ages. They cooperate to provide sustenance for the whole pride, to protect their young, and to protect their territory. They gang up to drive off other lion trespassers, both male and female. At night, lions, particularly the males, patrol their self-proclaimed borders, spraying urine on shrubbery to scent mark with their own unique signature and roaring mightily to warn other lions that trespassing will be met with attack by not just the large male but also the larger females of the coalition. Such teamwork is, of course, in the lions' own self-interest, since most of their food is gathered within the pride's territory.

A male lion may weigh up to 550 pounds, eat up to 60 pounds of meat in a single feeding, and sleep 20 hours per day. He requires a two-year apprenticeship to learn how to hunt. However, two-year-olds still have a lot to learn. Watching a young male and two of his sisters try to bring down a wildebeest, I witnessed their lack of experience. One female managed to clamp her jaws on the gnu's sensitive snout, while the male and female, attacking the rump, tried to knock the gnu over, but failed again and again. Meanwhile, the wildebeest seemed to be in a state of shock, not trying to fight back, because of either exhaustion or pain. (Such behavior is not unusual among prey animals; they seem to accept their fate.) Finally, they succeeded in toppling their prey, and the other female bit down on the throat, suffocating the doomed wildebeest. The belly was torn open and the male began to feed first.

Some males form lifelong alliances, usually with cousins and brothers.

Some large prides may have as many as eight adult males that maximize defense of the coalition. If the pride contains several males, usually only a few of them mate; but since male coalitions are made up of genetically closely related animals, genes common to all of the males are passed on by the mating males. Small prides usually have only a single male member. If he is old, he can be displaced by a younger, stronger, more vigorous male with a potentially long reproductive life. When this happens, the females with cubs may kill the cubs of the old male, or else the new king of the pride often kills the cubs. In either case, the females deprived of cubs quickly go into heat and mate with the new male—a process which selects for competitive genes.

Females are generally smaller than males; full-grown lionesses weigh up to 300 pounds. Usually both males and females hunt, but females tend to hunt more often than males. Sometimes, when hunting smaller antelopes, a female hunts singly while her sisters sit nearby, watching patiently; once the kill is made, they come up and share the spoils. When hunting larger prey, team action is the rule. The tactics vary with the type of prey. For elusive, agile prey, the pride may drive the hunted animal into an ambush; but for large, powerful prey, such as the dangerous, heavily horned Cape buffalo, the pride will isolate and surround their target, distracting the cornered buffalo until one lion can leap on its back and bite through its spinal column. The lion's most frequently used maneuver is a charge at the rear end and then a leap to grasp the rump with its clawed forepaws. Lions have very powerfully muscled shoulders and forelimbs and can turn their wrists inward to ensure a good grasp. Once the prey is down, a throat bite suffocates the victim.

Having dispatched the prey, the whole pride begins to feed, including the nursing mothers and the older cubs. There is a lot of squabbling, and when the male or males arrive, they claim the best feeding positions. When all are sated, they rest. The mothers with young cubs are the social core of the pride; and exhausted, they lie down and sleep while their young cubs nurse.

Hunting is strenuous and not without risk from the flying hooves of zebras, the tusks of hippos, or the hooves of giraffes. The hunt may take hours, and lions are very persistent in the hunt. Frequently they are interrupted by a clan of hyenas, which are equally persistent and dangerous adversaries. If the hyena clan can recruit enough members so that their combined weight exceeds the combined weight of the pride members at the kill, the pride will retreat. On the other hand, if the pride is large, they will take over kills made by hyenas. Hyenas and lions are natural enemies and will kill each other but not eat the kill. Set upon by a hyena, clan lionesses frequently climb to safety in trees, because hyenas can't climb. The big males are protectors of the pride and usually go after the large female leader of the hyena clan. The results are predictable: the lion almost always wins.

The Clan

Spotted hyenas are bizarre-looking creatures, repulsive and heavily-built, but equipped with long, doglike legs. They live cooperatively and hunt cooperatively to a greater degree than lions. Truly formidable nocturnal predators, they live gregariously in groups called clans, consisting of as many as 85 individuals, including males, females, and pups. The clans are ruled by a dominant female that can weigh close to 200 pounds. The female hyena produces large amounts of male hormone, which makes her external genitals look like those of a male; she is very aggressive, possibly as a result of testosterone, the male hormone, which primes the brain centers involved in aggression. The cheek teeth of the hyena are very large, and the jaw muscles powerful, allowing hyenas to crush even large bones, which are then swallowed and digested. The excess calcium is eliminated with the stool and the dried hyena droppings are white.

Hyenas' hunting strategy differs from that of lions. Whereas lions use stealth, ambush, and short-range charge, hyenas are pursuit predators capable of sustained chases despite their ungainly looking stride. They have a greater ratio of fatigue-resistant muscle fibers along with large, efficient hearts and lungs to oxygenate the muscles. With greater stamina than their prey, they are able to box in a herd and get it running, exhausting younger prey or older weakened prey that then begin to falter and are swarmed over. A frenzied mob of ravenous hyenas feeding on the kill is a noisy, nasty free-for-all. Higher social rank ensures first access. Animals of lower rank are threatened and attacked only by individuals above them in the hierarchy. All adult females outrank males, and rank is often conferred by the mother's status. If the leader of the clan produces two female offspring, one of the sisters will kill her sister shortly after they are born through repeated attacks of biting and shaking. Newborn hyenas have sharp teeth and high levels of testosterone and are fiercely aggressive. While such killing reduces the reproductive potential of the clan, it does select for a ferocious, aggressive leader.

During uninterrupted feeding, a clan can reduce a 300-pound antelope to a pile of shattered bones in a matter of minutes. Almost everything is consumed. Afterward, the mothers return to their home dens to nurse their young. Unlike lions, however, they do not nurse communally.

The Packs

Two canids (dogs) practice cooperative living, the wolves and the cape hunting dogs. The latter are the most efficient hunters of the African grasslands. The

The Leader of the Clan   Spotted hyenas live in social groups—clans—dominated by a single large female. While we tend to think of hyenas as carrion-eating scavengers, they are, in reality, awesome cooperative hunters and archenemies of the lion. Although smaller than lions, they can rout a lion pride when they outnumber their foes.

cape hunting dogs are quite distinct from any other canids. They are light-weight, long-legged, big-eared, and blotchy-colored. Possessed of incredible stamina, they can run down and exhaust most antelopes.

Before the hunt, there is little activity in the pack's den, which includes burrows for the young. The signal to get ready is given when the lead dog, a male, gets up, rousing the others into an energetic orgy of licking, grooming, and yapping. Then, the lead dog lopes off at a slow trot, with the other experienced adults following in a line. Once the prey herd is spotted, the dogs pick up the pace. Some flank the herd; others pursue from behind while the lead dog picks the target, which is usually a fatiguing youngster or a weakened older antelope. The pack now concentrates on one target, often with the lead dog's grabbing the tail to slow the prey down while the other dogs go for the under-belly. Once down, the still-kicking animal is disemboweled.

Unlike lions and hyenas, cape hunting dogs engage little in competition for food. Some may not even get to the carcass but wait around the fringes of the kill and beg for food, which a fully fed animal readily provides by regurgitation. When the pack returns to the den, the pups also solicit regurgitated food by begging.

The canid that has had the most impact on the human psyche is the wolf. Most likely, these large, social dogs were the ancestors of all domesticated dogs. They have a long mythology associated with them. In Europe, the ubiquitous wolves and their nocturnal, melodious howling inspired tales of werewolves, children's fables such as Little Red Riding Hood, and the orchestral musical piece by Serge Prokofiev, *Peter and the Wolf*. The founders of Rome, Romulus and Remus, were supposedly nursed by wolves, as was Mowgli of Kipling's *Jungle Book*. And, of course, there is the big bad wolf at the door of the three little pigs. Some horror stores about wolves were not without foundation: two centuries ago in Europe, wolves killed and ate many shepherds, shepherdesses, and children. The problem became so acute that Louis XIV of France had an entire military unit assigned to killing wolves. On the other hand, there are no documented wolf attacks on humans in North America.

Wolves, *Canis lupus,* are intelligent, curious, gregarious, temperamental, sometimes affectionate, melodious, communal dogs. The top predators in many food chains, they are tireless pursuers, seemingly able to glide across miles of terrain in search of prey. They are the ideal communal hunters, living in packs of up to a dozen animals. Each pack is composed of a dominant large male that may weigh up to 130 pounds (although in the arctic subspecies, males reach 175 pounds), a dominant smaller female, several juveniles, and pups. Social status within the group is communicated by posture, fang bearing, facial expressions, barks, and growls to establish dominance. In contrast, subdominant animals signify their place in the pecking order by giving a submission signal, namely lying on their backs and exposing their throats. Such gestures almost always stop aggression. The pack is very territorial, scent marking its boundaries with urine and announcing its turf at night with choruses of haunting and melodious howls that can be heard miles away.

Pack life involves a lot of play, tussling, and stretching prior to setting off on the hunt, which is led by the dominant male and female and the older, experienced pack members. The juveniles and nursing mothers remain behind, guarding the pups. When hunting as a pack, wolves go after prey much larger than themselves: bison, deer, caribou, elk, musk oxen, wild sheep, and mountain goats. (Unfortunately, they also, sometimes, prey on domesticated animals, thereby coming into direct conflict with the most potent predator, man.) The attack is precise and relentless. First they try to get the herd running; then they work to split the herd. Then, pushing the fleeing herbivores to exhaustion, they will pick out a young or weakened animal, gang up on it, and make the kill, which is shared by all members of the pack. In order to sustain itself, an adult wolf must eat the equivalent of 18 to 20 deer per year.

Wolves hunt small prey individually, favoring hares and voles. The arctic subspecies, *Canis lupus arctos,* is one of the toughest, most ecologically elastic predators. Its hunting ground is one of the world's most inhospitable terrains, where in winter, the temperature seldom gets above −20°F. Despite being targeted by humans and climatic extremes, the wolf still survives. Moreover, having earned the grudging respect of humans, it is now being protected.

Solitary Carnivores

The majority of the mammalian carnivores are solitary hunters that socialize with members of their own species only during the mating period. The largest of these terrestrial hunters are the polar bears, whose ancestors were already established in Europe 20 million years ago in the early Miocene epoch. About 200,000 years ago these large, brown, furry hunters began to exploit the ice caps of the Arctic and ascended to the role of a top predator.

An adult male polar bear, *Ursus maritimus,* weighs over 800 pounds. There are reliable records of 1,600-pound individuals; this makes the polar bear the largest modern terrestrial carnivore. They have typical canid teeth, their massive wedge-shaped heads are almost 2 feet long, and their 10- to 12-inch-wide clawed paws are lethal weapons. The paws may be used to batter a prey: a single swipe can kill a seal. When the bears swim, they use their forepaws as paddles and their rear legs as rudders. Extraordinary swimmers, polar bears often submerge and swim under the ice to get close to their favorite prey, which are seals, without being detected. They are patient stalkers with excellent long-range vision for target acquisition. However, sometimes their prey spots them before they get close enough to charge and dives to safety. But seals, being air-breathing mammals, have to come up for air, and the ice cap is dotted with breathing holes. So, the polar bear shifts tactics and waits silently and patiently to ambush a seal. When it comes up for air, the bear dispatches it quickly with a bite to the head. Polar bears also use breathing holes to attack basking seals at the hole's edge. They swim under water, popping their head up through a hole and scanning the surroundings. Then, moving from hole to hole, they finally reach their unsuspecting target, explode up from underneath, and make the kill.

The prey is heavily larded. The bear concentrates on eating the high-calorie fat, often leaving the remainder of the carcass. Since they know of this feeding strategy, arctic foxes have learned to follow polar bears to scavenge the remnants of the kill. During the hunting season the polar bears eat prodigiously, packing on up to 5 inches of insulating fat before going on a prolonged fast when the ice floes begin to break up.

The ancestral bears that gave rise to the polar bears produced another line of giant bears, the great cave bears that were contemporaries of early European humans. In response to drastic upheavals of nature, such as extreme climatic change, these now-extinct carnivores adjusted by changing their feeding habits: they became generalists, adapting to a great variety of feeding options. From these opportunistic generalists, modern bears evolved, keeping their options open and thus ensuring their survival. Of all the bears that catch the human imagination, the grizzly bear, *Ursus arctos horribilis,* reigns supreme. Large, heavy, muscular, and covered with brown, shaggy fur, it is indeed a fearsome predator, capable of bringing down a large moose or elk. It derives its name from the silver-tipped (or grizzly) hair on its shoulders and back. Grizzly bears are found mainly in northern parts of North America, and their size varies considerably. The largest, the Kodiak grizzly, is almost as heavy as the polar bear.

During the spring and summer, after emerging from their long winter torpor, these omnivorous browsers feed on a wide variety of foods: berries, fruits, roots, bees' nests, nuts, and even carrion. Being the penultimate generalists, their diet includes rodents and other small burrowing mammals which they dig up with their extremely long, nonretractable claws at the front of their broad flat paws. The forepaws are used not only for digging but also for fishing and for grasping large prey. Fishing tactics vary. Sometimes a quick bite can even pick off a leaping salmon in midair, or they may flip the fish onto the shore with a well-aimed swipe of their forepaw. When the salmon are running, the grizzlies, given the abundance of food, become very picky eaters, often eating only the fatty skin and leaving the remains to the multitude of waiting scavengers.

While not particularly territorial, grizzlies turn ferocious when threatened, and a single blow from the bludgeon like forepaw can kill an adult human. Although rare, grizzly attacks on humans have occurred, and contrary to popular belief, climbing a tree to escape is not a viable option: these bears can climb.

The Solitary Cats

Today there are about 37 known wild species of cats. They range in size from the typical domestic cat to the true king of the cat family, the Siberian tiger. These huge felids should not be confused with the prehistoric sabre-toothed tigers, since they represent a totally different branch in the evolution of cats. The sabre-toothed cats were contemporaries of early humans. With their enormously long canine teeth, they preyed on large, thick-skinned herbivores like the wooly mammoths and became extinct less than a million years ago. Modern tigers, to the best of our knowledge, evolved from a stem ancestor in South China slightly more than 2 million years ago. These cats radiated over much of Asia, some as far eastward as the Caspian Sea. They were initially classified as *Felis tigris,* but subsequent anatomical studies revealed that tigers and their cousins, the lions, the leopards, and the jaguars, had elastic bony supports for their tongues, so they were reclassified into the genus *Panthera*. The combination of these elastic bones and the peculiar structure of their vocal apparatus (larynx) acts like a slide trombone to produce the deep roars that carry over long distances. These same structures also allow them to purr when exhaling, while the smaller cats purr during both inhalation and exhalation.

Tigers

Tigers are powerful, solitary predators, superbly armed to kill prey larger than themselves, although they will eat anything that presents itself. They arc stealthy stalkers whose dark stripes act as a disruptive camouflage preventing potential prey animals from recognizing their distinctive shape. Being so endowed, tigers blend in with their background, crouching close to the ground in ambush. Their forward-facing eyes give them a three-dimensional color image of their target.

When hunting at night, tigers have a tremendous advantage, since they have many rods in their retinas, more than humans. Behind their retinas is a reflecting structure that more than doubles their ability to see in dim light. This reflecting layer accounts for the eye shine seen when a beam of light hits the tiger's eye. Many nocturnal predators, including hyenas, lions, and croco-

The Stealth and Camouflage of an Ambush Hunter   The Bengal tiger, top predator of India's forests, is a master of stealth. Blending in with its surroundings, it waits motionless, muscles taut, and—once its prey is within range—bursts from its concealment in a high-speed charge.

diles, produce these eerie glowing eye shines, but somehow tigers are singularly recognized for them, as in "Tiger, tiger, burning bright." The target information gained from their eyes is augmented by their acute sense of hearing, and their ears are optically aimed in the direction of the prey as well. So, both auditory and visual data are processed simultaneously.

The commands relayed to the tigers' motor control centers result in a sudden charge or leap, driven by the thrust of its exceptionally powerful rear legs.

This animal's leaping ability is truly amazing: there is a recorded case of a tiger pulling a man out of a tree in which he was 18 feet above the ground. As the tiger leaps, the strong, sharp, retracted claws are unsheathed from their scabbards. The tiger then twists its flexible wrists inward to grasp the prey. The bite, delivered to the nape of the prey's neck, is sufficiently strong to kill small prey. Larger prey are wrestled to the ground, and then the tiger shifts its bite to the throat, killing the animal by suffocation. The carcass is usually dragged to cover and feasted on. The tiger can consume up to 100 pounds of flesh in a single sitting. Afterward, it covers the carcass with leaves and rests before returning for another meal.

The tiger's sense of smell, while good, is not particularly important in tracking prey but is used instead as a receiver in communication to identify marked territory, readiness for mating, and possibly even another tiger's age. Tigers mark their territories with piles of unburied droppings. Frequently, tigers stand on their hind paws and claw trees, leaving scratch marks to announce not only their presence but also their size. The tiger's perfume can even be detected by humans. When sniffing scent markers, tigers and many other animals seem to grimace hanging out their tongues, curling their lips, and wrinkling their noses. This phenomenon, called *flehmen* by German naturalists, is used to bring odors to the Jacobson's organ, a taste-smell sensor in the roof of the mouth.

Tigers vary considerably in size depending on their race, but males are always larger than females. The largest, the Siberian tigers, usually weigh from 400 to 600 pounds, but the largest one on record weighed a whopping 850 pounds. These huge, solitary hunters cover hundreds of miles in search of prey. As winter approaches, the tiger's coat becomes longer and thicker; its orange stripes fade to help conceal it in its frigid, snowy habitat; and it packs on insulating fat. Unfortunately, due to human hunting and habitat destruction, all remaining races of tigers are considered endangered species in the wild.

There are a number of other large and midsized solitary cats whose hunting methods are basically similar to the stalk-ambush mode practiced by lions and tigers. One member of *Panthera* is different in that it is a chaser, a cursorial speedster that holds the reputation as the fastest land animal. It is the cheetah.

Cheetahs

Cheetahs derive their name from the Hindi word *chita* which means "spotted one." Just one glance at one of these lean, lithe, spotted cats gives the impression that they are built for speed. Although there is some debate about how fast the cheetah can go at maximum speed, the consensus is somewhere between 40 and 70 miles per hour. It can shift from a slow trot to maximum speed in about 3 seconds, and high-speed cinematography shows that as it bounds forward, it goes airborne, covering 20 feet with a single bound. To attain such speeds, its leg bones and associated musculature are attached to a flexible skeletal framework in which the hip and shoulder girdle are loosely attached and can swivel, allowing maximum extension. At one point in the pursuit, the cheetah is airborne with all four legs bunched under it and its supple backbone arched like a bow pulled taut. The stretched elastic tendons, attached to vertebrae, rebound as the muscles straighten the back, literally shooting the cheetah forward. As it lands, it plants its extremely long legs and leaps once more into the air. The cheetah's feet have short, nonretractable claws, which give it good ground contact when running at full speed. However, the propulsion is provided by fast-twitch muscle fibers that rapidly fatigue, so the cheetah can keep up the chase for only a few hundred yards before it is exhausted.

Its prey, small antelope, are also speedy. Some, like the impala, are capable of prodigious leaps and don't make easy targets. If the cheetah succeeds in catching up with its prey, it knocks it over and lunges for the soft underside of the prey's throat. Grasping the antelope's windpipe in a viselike grip, the cheetah dispatches it by suffocation. The cheetah's precision killing tool, unlike that of the bigger carnivorous cats, is its short muzzle, designed for gripping rather than tearing.

Cheetahs are visual, daylight hunters of the open grasslands, and their large, forward-facing eyes provide them with maximum binocular color vision. Prior to the hunt, they frequently perch on a fallen tree or termite mound, so they can view their prey from a height. We had a family of cheetahs near our

Evasive Action in Escape   Flat-out speed will often take prey animals out of harm's way, but sometimes speed alone is insufficient. To enhance their odds of survival, many prey animals (such as the impalas shown here) take a variety of evasive actions, including spectacular leaps.

farm in Kenya, and they frequently climbed up on the roof of our Volkswagen van to peruse the area. At other times they would rest in the shadow of the van. We followed a mother and three cubs for almost a year, and they were friendly and apparently unafraid of us. As a matter of fact, cheetahs have a long history, over 5,000 years, as domestic pets and trained hunters.

The young undergo a long apprenticeship, up to 20 months, under their mother's tutelage to gain the skills needed to be efficient hunters. Eventually the young adults leave the mother, and often brothers from the same litter form hunting groups that stay together for life and share their kills with each other, but hunt individually. Since they are easily driven away from their kill by hyenas and lions, they hunt frequently. Lions actively hunt cheetah dens and kill the cubs, but don't eat them. The reason for this behavior is unknown.

There are other mammalian predators of the forest and open grasslands that are not carnivores; as a matter of fact, most of the time they are omnivorous gatherers that kill prey only occasionally. Although they lack the speed, size, and weaponry of the dogs and cats, they are related to the ultimate predator, man. These animals are the primates (*primata* meaning "first ones"). They alone have ascended to occupy the top position in the evolutionary ladder. Two groups of primates are known to hunt other animals. These are the baboons and the chimpanzees; the latter are genetically closer to humans than any other primate.

Trooping

At some crucial stage of evolutionary development, some primates came down from the trees and took up part-time residence on the ground, thereby widening their choice of foods. Among these were the baboons, whose doglike, long faces and forward-facing eyes give them an uninterrupted view of their surroundings. Baboons live in highly structured societies called troops that are managed by social hierarchies. The male hierarchy is dominated by one individual, the so-called alpha male. He earned this status by fighting his way to the top, and he brooks no challenge from any of the lower-ranked males. Other big males close to him in the pecking order are careful to avoid getting too close to him, lest they become targets of his wrath. Lower-status males sometimes get

Members of the Troop (facing page)   Baboons are social animals that live in highly structured troops dominated by a single large male. Male baboons possess long snouts and long fangs, and while they are opportunistic hunters, their diet is mainly vegetarian. When a troop of baboons is attacked by a predator (for example, a leopard), its large males form a defensive phalanx to protect its females and younger members.

close, but they give submissive gestures acknowledging his high status. This status is evident in his behavior, his odor, his brain chemistry, and his hormone levels. When one of the younger males decides it's time to challenge the alpha male, the imminent shift is somehow sensed by the females, some of which start cozying up to the challenger.

Females and juveniles also establish social rank hierarchies. Not even the highest-ranked female, however, will stand up to the lowest-ranked male. When the troop travels it has a more-or-less standard formation with subordinate adult males to the front and rear and with the dominant male, females, juveniles, and infants in the center. When danger, for example, a leopard, threatens, all long-fanged males rush to the fore, presenting a phalanx of defenders, while the rest of the troop scurries to safety. Only the males have long, sharp canine teeth, and these are awesome weapons, used to maintain status and to fight off animals that prey on the troop. Although baboons use all four limbs for walking, they have the ability to rotate their thumb so that it can be held opposite each of the four fingers. This ability was the forerunner of tool usage.

Sometimes male baboons are opportunistic killers. Should they come across a young baby antelope, they pounce on it, grabbing it with their forepaws, and kill it with a bite or by smashing it on the ground. This behavior was the beginning of the gatherer-hunter mode of existence. One primate, however, went beyond the purely opportunist form of hunting; it became a true hunter and a very efficient one. That primate was the chimpanzee.

Chimpanzees

Until the pioneering studies of Jane Goodall, we tended to view our closest living relatives as relatively benign, peaceful herbivores, intelligent and amusing animals. But Goodall's studies and other more recent research found them to be highly efficient, cooperative predators. Chimpanzees live in tight-knit communities of up to 75 individuals led by a dominant male. Laboratory studies of captive chimps have shown them to be highly intelligent, capable of solving problems in novel ways, and endowed with some capacity to use tools. Their use of long stems of grass and slender branches stripped of their leaves to probe into holes in termite nests to extract these delicacies is well documented. They are also known to chew leaves and use them as sponges to sop up drinking water. Among other natural tools are sticks and stones to crack open hard-shelled nuts; this hammering skill is taught to young chimps by their mothers. Furthermore, to facilitate nut cracking, chimpanzees leave the hammering stones around the bases of nut-bearing trees. One of the hallmarks of chimpanzee evolution is the sharing of food with other members of the troop, which includes the sharing of meat from kills.

Male chimpanzees are powerful creatures, some weighing over 90 pounds. Despite this bulk, they are excellent climbers. Their favorite prey are the somewhat slow, red colobus monkeys which when full-grown may weigh 30 pounds. Chimps seldom go after the faster, more agile types of monkeys. The dominant male, whose superiority is uncontested, leads the monkey hunt, and excitement builds in the whole troop. With a number of experienced male hunters, the leader sets off into the woods, scanning the forest canopy for prey. The approach is silent. Flankers race along the sides of the quarry's path, while others race ahead to block off escape routes. With the prey encircled, one or two chimps climb into the canopy in pursuit. Finally, when the chimp catches up with the colobus, it grabs it and kills it with a neck bite. When hunting cooperatively like this, chimpanzees achieve a 50 percent efficiency rate which is higher than that of any predatory cat or dog cooperative hunter. Chimps also use cooperative tactics to hunt young antelopes and wild pigs.

Once the kill is made, the prey is torn apart and shared with the troop, who seem to be celebrating the victory by hooting and drumming on trees. This drama, acted out over and over again, represents the culmination of group hunting practices by nonhuman animals. It could indicate that our early hominid ancestors used strategies and tactics like those used by present-day chimpanzees.

Chapter 5

The Wild Blue Yonder

The atmosphere, or air space, was the last environmental niche to be occupied by living things. The first true fliers evolved from ancestral insects which had already occupied aquatic and terrestrial environments. About 210 million years ago, reptiles became airborne, first producing long-tailed, small fliers, and later, the largest of all flying animals, the pterosaurs. Among the pterosaurs was a giant with a 40-foot wing span.

All the flying reptiles and their dinosaur cousins were wiped out during the great Cretaceous-Tertiary extinction. However, from the ancestral reptile stock, modern birds evolved, diversified, and now range the skies from pole to pole and from sea level up to altitudes of 20,000 feet. The last animal to acquire a capacity for flight was the mammal known as the bat, whose flying skills are extraordinary. Finally, humans, the toolmakers, achieved powered flight only a hundred years ago, and although human-designed flying machines are awesome, the aerodynamics and flight capabilities of insects, birds, and bats have yet to be equaled by humans.

Mastering the Aerial Environment

All organisms that overcame the force of gravity and entered the three-dimensional environment of our atmosphere had to deal with four basic factors: the weight of the flying machine, including its fuel; lift, the force to overcome gravity; drag, a retarding pull exerted by the air; air currents moving up, down, and sideways; and finally, thrust, a force that allows the flying machine to overcome drag and air currents and drives the flying organism through the air. Animals that fly have dealt with the problem of lift by evolving wings. Airflow over the wings creates an imbalance of pressure, wherein the air pressure below the wings is greater than the air pressure above the wings. As the wings are forced upward, they carry the body along with it. In order to overcome drag or rearward thrust and air currents, flying organisms had to develop sufficient thrust to overcome drag.

In addition to acquiring the aerodynamics, which were needed for flight mechanisms, flying animals evolved to ensure stability while airborne. The three major disruptive components encountered in flight are pitch (head up or down), yaw (slipping sideways), and roll when its wings dip to one side or the other, causing the animal to roll along its longitudinal axis. We will examine how each of the groups of flying animals dealt with these problems, but first we'll make a digression to examine gliding and soaring.

Gliding, Soaring, and Wafting

The capacity to glide was the first development in the evolution of living flying machines. Gliders use lift provided by winglike structures and the force of gravity to fly through the air. They lack any thrust-producing capability but can maintain airborne stability. For this reason, gliding is always of limited duration and range. A number of misnamed animals are competent gliders. For example, when escaping from predators, flying fish can produce sufficient forward thrust in the water by contracting their muscular tails sideways so that their forward parts slant upward. This position allows them to spread their outsized, membranous pectoral fins, which act as air foils, permitting the fish to glide for several seconds. Flying fish may also be using an air cushion produced by skimming low over the water, a technique used by flyers returning from overwater flights low on fuel to maximize their range. Other so-called flying animals, squirrels, snakes, and lizards, use gravity's pull and distensible membranous appendages to convert a free fall from a height into a controlled glide of short duration.

True flying animals also glide but have the capacity to recognize and use rising columns of heated air, updrafts called thermals, to lift them to higher altitudes as well. The use of thermal currents is called soaring, and many birds of prey, such as eagles and hawks, are experts in this skill. The undeniable master of soaring is the Andean condor whose 10- to 12-foot wing span can carry it to altitudes of over 20,000 feet. The great benefit of soaring is that it requires very little energy for muscular contraction of the wings. Some large predatory seabirds, such as the long-winged albatrosses, can soar for hours without flapping their wings.

One other group of animals can be found airborne by the thousands at altitudes of 200 to 16,000 feet. They are spiders who sail through the air on "gossamer wings" of silk. Spider flight, or ballooning, is practiced by the young of a number of species of spiders. When the Indonesian volcanic island of Krakatoa exploded, all life on the island was wiped out. Scientists studying the repopulation of the island found spiders among the first arrivals. To achieve flight, young spiders climb on a leaf; assume a head-down, tail-up stance; and begin to secrete webbing. The webbing is picked up by any passing breeze, and the spider, attached, sails off downwind. To land, the spider simply rolls up the silk, decreases the gossamer sail's lift, and descends at will.

The First True Fliers

The pioneers of powered animal flight were those insects which became airborne over 300 million years ago. They evolved a very lightweight airframe, composed of a strong, jointed, external skeleton, to which the flight muscles of the wings are attached. The motion of the wings is in the shape of a looping figure eight. Thus, the powerful downstrokes create both lift and thrust, and drag is minimized because the wings are tethered during the upstroke. We will elaborate on the extraordinary aerodynamics of insect wings later, but it should be stressed that insects are more like helicopters than fixed-wing aircraft, and many flying insects can hover, rise up or down almost vertically, and even fly backward.

Wing beating in insects is much faster than in other flying animals. (The frenetic oscillations of bee wings, up to 300 beats per second, were captured musically in Rimsky-Korsakov's *The Flight of the Bumblebee*.) Some smaller insects, the miges, have wing-beat frequencies as high as 1,000 per second. With wing beats of such frequencies, you would expect some to move at very high velocities; but the fastest insects, the dragonflies and darters, can only manage about 40 miles per hour during short-distance dashes, and their cruising speeds are considerably slower. Most predatory insects do not hunt other airborne moving targets, but those that do are extraordinary aerialists, marvelously adapted for visual target acquisition, pursuit, and intercept operation.

Aerial Daytime Interceptors

Patrolling the daytime air over a quiet pond, flies a brilliant, iridescent, blue-green, voracious predator, *Anax Junius,* the "lord and master of June." This fierce aerialist belongs to a group of insects called the Odonata, which means "toothed ones," or more commonly, dragonflies. These speedy flyers can take off backward, hover, accelerate in a fraction of a second, make an unbanked turn at full speed as if on a swivel, somersault, and even stop on a dime in flight. Although their wing beat is relatively slow, 30 to 50 beats per second, each beat is extremely powerful since almost half their weight is made up of flight muscles.

Over the span of 300 million years, dragonflies have had ample time to perfect feats of aeronautical engineering that are truly amazing. They have four

The Lord and Master of June's Skies Manfred von Richtoven, the "Red Baron" of Germany, was the most successful aerial hunter of World War I, scoring some 80 kills during his short career. Yet, a dragonfly, in one day of hunting, can score up to 400 kills by using its extraordinary flying skills and remarkable multifaceted, bulbous eyes to zero in on its prey.

wings. Each wing is a transparent sheet of intricately cross-braced struts forming a complex mosaic of trapezoids, pentagons, hexagons, and octagons. Blood flowing through the thicker struts keeps the airfoils stiff.

The dragonfly can automatically bend its wings to produce a multitude of aerodynamic effects. It can use all four wings in unison or beat the front and rear pair out of phase while maneuvering in flight. The upper wing surfaces are richly endowed with sensory hairs which work together with sensors on their stubby antennae to detect tiny changes in airflow. On the wings' leading edges are blood-filled thickened structures, called pterostigmatas, that seem to act as stabilizers. The dragonflies generate turbulence on the dorsal wing surface to enhance uplift; this action is something our modern aeronautical engineers can't duplicate.

An Aerial Hunter Becomes the Hunted While dragonflies are the deadliest hunters around ponds and swamps, they too are hunted despite their flying skill and 360-degree field of vision. The bullfrog shown here has just caught a dragonfly in midair. Sitting motionless and well camouflaged (because its color matches its surroundings), the bullfrog waits patiently for its target to get within range. Its incredible eyes track the target and relay relevant data to the bullfrog's brain, from which commands are sent to its long, prehensile tongue, which behaves like an optically aimed, wire-guided missile. The end of its tongue is sticky and the dragonfly, adhering to it, is pulled into the bullfrog's mouth.

Not only can dragonflies outmaneuver any flying insect; they have better equipment for locating and tracking their prey. Their bulbous heads are dominated by two huge eyes that give them a visual field covering almost 360 degrees. Each eye is composed of a mosaic of sensory structures called ommatidia. Each ommatidium has its own lens and nerve supply which relay information regarding size, speed, and direction to the insect's on-board microprocessor, its brain. This information allows for flight corrections the insect needs to complete interception. We humans have developed phased-array radar that do somewhat the same thing. However, our version is not miniaturized, nor is it as effective as the dragonflies' visual processor.

Prey are caught in flight by the dragonfly's hooked legs. Each of the six legs has a row of very sharp spikes that terminate in a two-pronged, clawlike structure. All six legs may be used in the catch, and then the front legs transfer dinner to the mouth. While the dragonfly's usual tactic is to eat smaller prey like mosquitoes and flies in flight, it most often lands when it aims to devour larger prey—bees, moths, and even other dragonflies.

Despite its skills as an aerial hunter, most of the dragonfly's predatory life is spent underwater as a bottom-dwelling larva, undergoing growth and development. In this aquatic habitat, the larva is the scourge of the scummy bottoms, where it hunts and eats everything, including most other aquatic predators. The larval dragonfly's unique weapon system is a prehensile, toothed lower lip that it can shoot out like an optically aimed wire-guided missile to hook its prey.

However, even predators sometimes become prey to something bigger (as a matter of fact, dragonfly larvae make great bait for fishing). But the larvae have a secret weapon. Their gills are in their abdomen, and their muscular rectum is used to keep a flow of oxygen-rich water over the gills. When under attack, they can forcefully contract the rectal muscles and rocket away from the predator on a jet of water. While such forceful water flatulence may seem unrefined, it does enhance survival.

Winged Ogres of Ambush

For sheer ferocity, no other winged insects can match the mantids. Although they can fly, they don't catch their prey while on the wing. The famous French naturalist Fabre called the mantid "the ogre in ambush that demands a tribute of living flesh." Mantids are masters of stealth, deception, and camouflage, capable of lying in wait, motionless, until their target is in range and then striking out like a lightning bolt to skewer their prey on their saw-toothed front legs. Stealth is accomplished by camouflage. Mantids can match the green of foliage or the brown of leaf litter; some are flower-like, lichen-like, barklike, or twiglike, depending on their hunting environment. They were around long before birds and mammals appeared, and their 1,800 species inhabit all regions of the earth with climates suitable to their survival.

The outstanding characteristic of these rapacious predators is their astonishing weapon system, consisting of their scimitar-like forelegs. Carolus Linnaeus, the Swedish naturalist, gave the European mantis the name *Mantis religiosa,* meaning "praying mantis." While at rest, it carries its weight on its pairs of middle and hind legs, while holding its sharply spined forelegs folded in a prayer-like posture.

Winged Ogre of Ambush   The praying mantis selects a suitable spot from which to ambush its prey and changes its color to blend in with the surroundings. Once in position, it waits motionless, its raptorial-toothed forelimbs held forward as if in prayer, for a meal to come within range. The strike of these forelimbs is optically aimed and lightning-fast. Once in the mantid's grasp, the prey is presented to a surprisingly small mouth and leisurely consumed.

These forelegs are traplike appendages, greatly elongated and robust and held like a half-opened jackknife, as if in supplication to a divine being.

The inner faces of the forelegs are grooved, ridged, and armed with rows of small and large spines; at the distal tip of each foreleg is a hooked, pointed claw. Mantids appear to have an instinct for positioning themselves where prey traffic is heaviest. The strike of the forelegs is optically aimed and very fast, a twentieth of a second.

The optical aiming system is truly remarkable. The head of the mantid is loaded with optical sensors, thousands of ommatidia assembled in a pair of compound eyes. The eyes are locked in a forward position, and each eye has a dark, dotlike porthole that gives the mantid the visage of death, the final apparition seen by the prey just as the forelegs clamp down on it. While the eyes are frontally locked, the head is amazingly mobile: as it swivels, it activates

Using a Threat as a Deterrent   Mantids, although awesome predators in their own right, are often hunted by a variety of other predators. When confronted, a mantid rears up, exposing a pair of false eyespots, and spreads its wings to make itself appear larger. Such displays sometimes work, allowing the mantid to return to its "hunter mode."

sensory hairs, enabling the mantid to triangulate on its potential victim. Once the prey is in the grasp of its forelegs, the mantid brings its victim to its surprisingly small mouth and chews it up.

The mantid will eat anything, even prey larger than itself, including other mantids—brothers, sisters, parents, mates. Sometimes while copulating with the smaller male, the female will bite his head off. This tactic would seem to be counterproductive, but the part of the brain containing the copulatory center is in the mantid's thorax, so the headless male continues to copulate. Then the female finishes off the rest of the meal. This is one case of truly losing your head over a woman.

The cannibalistic mantids are prey to many other predators besides their own kind; birds and small mammals apparently find them delicious. When pursued, their awkward, slow gait and equally ungainly, slow flight make them easy targets for insectivorous (insect-eating) predators such as echolocating bats. Some species of mantids, however, evolved an early warning system that can detect the bats' hunting pulses at distances considerably beyond the bats' echolocating range. This warning system gives the mantid time to seek cover. Many mantids, when confronted by predators, attempt to deter them by rearing up on their back legs, spreading their wings and forelegs to give an appearance of greater size, and flashing false eyespots. This last tactic has proven especially effective. Many predators are deterred and startled by false eyes on their prey. Consequently, false eyes evolved in many diverse species of insects and other organisms.

Blood-Eating Winged Predators

Predators that kill to obtain their nutrients automatically reduce the population size of their resources. Indeed, studies of the population dynamics of predators and prey show that when prey populations increase in size and availability, the predator population soon begins to increase in size. As the predators increase in number, they deplete the number of prey. As a result, the predator population, competing for a limited resource, declines. A better hunting strategy, it seems, would be one in which the predator periodically took only small samples of nutrients from the prey, leaving the prey population a stable, reusable resource. This type of strategy has, in fact, been exploited by a large variety of small predators whose prey is often several thousand times their size. These are the blood eaters.

Goethe, the great German poet and philosopher, referred to blood as "that wonderful juice." For blood eaters, that juice is abundant, high in calories and protein, and readily accessible. Moreover, it is a self-renewing resource. The only large blood eaters are humans. The nomadic cattle herders of the East African savannas, the Masai, often pierce the neck veins of their cattle to collect blood, which they mix with milk. When teaching in East Africa, I was invited to share in this nutritious drink, but I found it unappetizing and alien to my cultural background.

The most famous of the blood eaters are the vampire bats of South and Central America. As the whole idea of blood eaters has somehow become indelibly imbedded in the human psyche, any newly discovered blood eater is often given a common name in which *vampire* is a prefix. For example, the pencil-thin, parasitic fish of the Amazon basin that normally latches itself to the gills of fish, produces a wound, and eats the blood is called the vampire fish. This nasty little blood eater, attracted to the excretory products in human urine, has been known to swim up the urethra, lodge itself, and feed there. Also

Night-Hunting, Echolocating, Thermal-Sensing Blood-Eaters Vampire bats are nocturnal hunters and they locate their prey by echolocation—an out-of-water sonar system. Instead of landing *on* their prey, however, they land nearby and stealthily approach it on foot. Once "aboard" the prey, they use heat-sensitive receptors on their snouts to find hot blood vessels near the surface of the prey's skin. Then, using their razor-sharp teeth, they expose these blood vessels, nick them, and begin to drink.

in the Maylasian jungles, there is a nocturnal blood-eating moth whose piercing-sucking mouth parts can penetrate up to a half-inch into the skin; it is appropriately called the vampire moth. Since people name organisms, it is not surprising that one of the predatory vibrio bacteria is called vampirobacter.

There are several species of bats that eat the blood of warm-blooded animals—fowl, cattle, and sometimes humans. On moonless nights, these echolocating, gregarious nocturnal hunters foray from their homes in hollow tree trunks and begin their patrol. Their primary system of target acquisition is echolocation. They seldom land on their prey, but instead land near it and hop, walk, and crawl onto the animal, which is usually drowsing or sleeping and so is less sensitive to the stealthy movements of the vampire.

At this point in the hunt, the vampire brings into play a unique sensory weapon system, similar to that used by some night-hunting snakes: thermal sensing. Two pits on the front of the vampires' face contain sensors that can detect tiny changes in temperature, changes that are produced by hot blood vessels near the prey's skin surface. Once having located a blood source, the bat, using its razor-sharp triangular teeth, begins to delicately shave away the tissue overlying the blood vessel, and then nicks the blood vessel and secretes its saliva. In this saliva is a potent anticoagulant, which ensures that the blood will flow. Also in the saliva is a vasodilator which increases the diameter of the nicked blood vessel. The bat then engorges itself with the prey's blood. Its feeding orgy lasts only a few minutes, but in that time the vampire can consume up to 40 percent of its weight in the host's blood. Blood eating is important to pregnant female vampire bats, particularly the most common species, *Desmodus rotundus,* whose protein requirements are high.

Unfortunately, vampires can, at times, carry the deadly rabies virus and infect the animals it feeds on. In South and Central America, hundreds of thousands of cattle and domestic fowl die from vampire-transmitted rabies. Hungry vampires will also feed on humans; in 1936 in Costa Rica, 89 people died from paralytic rabies. Today, vampire bats have a minimal effect on human health, but, if we judge from current movies and television shows, they still have an enormous impact on our psyche.

Blood-Sucking Insects

The most successful aerial hunters that feed on blood are insects belonging to the order Diptera, meaning two wings. They all have three basic body parts. First is a head, loaded with an array of extraordinary target acquisition sensors and piercing-sucking mouth parts. Second is a midportion, the thorax, which has three pairs of jointed legs; a pair of membranous, transparent wings; and a pair of knoblike wings that act as balancing organs. The third part is a segmented abdomen. The diptera are widespread throughout the world and are prodigious hunters. Unfortunately, they also act as vectors for a number of disease-causing microorganisms.

Some of the most voracious of these tiny-winged blood suckers are the humpbacked, stout-bodied black flies, or turkey gnats. They breed in running water. When the adults emerge, they swarm up in prodigious numbers, attacking en masse any exposed surface. The other blood-sucking diptera include the night-hunting biting midges and gnats called "no-seeums" that can get through the finest screening. The weak-flying nocturnal hunters called sand flies can make camping out a real nightmare. Much could be written about these small diptera, but the real arch villains are their cousins, the mosquitoes.

There are over 3,000 species of mosquitoes. Only a small portion of them feed on the blood of mammals, birds, and reptiles. Blood-sucking is performed only by females and only when they need extra protein to make their eggs. Unfortunately, some of these blood-sucking species seek out and feed on humans. In so doing, they act as vectors for a number of deadly viral, bacterial, and parasitic diseases.

Mosquitoes begin their lives as eggs laid in freshwater. After passing through larvae and pupae stages, adults emerge, and within a day, the young adults begin to seek food with a remarkable array of sensors. Among these sensors is a pair of multifaceted compound eyes capable of visually tracking a moving target. However, the mosquitoes' primary target acquisition mode is odor. The slender antennae are loaded with sensors that detect specific chemicals on and emanating from the host. These include skin oils, lactic acid (a metabolite), adenosine triphosphate, amino acids, sugars, and possibly even sex hormones. Hungry females are also able to sense carbon dioxide concentration in the air. They prefer warm, dry target surfaces to cold, moist ones.

Within a few seconds of detecting its prey, the mosquito makes a direct approach, lands, and begins to probe the skin with its proboscis, which contains sensors (palps) and an array of other structures, including a needle-sharp stylet. With all six legs firmly on the skin, she uses a muscle-driven saw-toothed component of the stylet to penetrate the skin of the victim, a process that takes almost a minute as she probes around to find a rich blood supply. If her initial search is unrewarding, she withdraws her proboscis and tries again. There is a salivary canal in the stylet, and the secreted saliva acts both as a vasodilator, increasing blood flow, and as an anticoagulant. Feeding takes a little over 2 minutes. During this time, the mosquito ingests about 3 milligrams of blood (about one-millionth of an average human's weight).

The blood sucking is powered by muscular pumps on the anterior digestive tract. As the digestive tract fills, the abdomen stretches and, in so doing, activates stretch receptors that send neural signals to the mosquito's brain saying, "Tank is full! Stop pumping." During this sucking phase, she consumes two to three times her own weight. She is now so bloated and heavy, she can barely fly. She then looks for a quiet place to digest her meal. Like any binge drinker, the first thing she does is excrete water (blood is 55 percent water). In a few hours, she can begin to digest the supernutritious remnants of her dinner. She may not feed on blood again, because a built-in control system turns off the hunting drive. After all, hunting is dangerous, and during the feeding stage these delicate hunters are easily swatted into oblivion. For this reason, a well-fed female mosquito switches her diet to sugar-rich plant juices.

Not all predatory species of female mosquitoes follow this game plan. There are a number of variations on the method presented here, which is that used by *Aedes aegypti*, a handsomely striped, deadly flying tiger that transmits a number of incapacitating and lethal diseases.

Wingless Aerial Hunters

Fleas are thought to have evolved from some ancestral insect about 60 million years ago. To date, more than 2,400 species have been catalogued. Most of them prey on warm-blooded mammals, large and small. They range from the Arctic Circle to Antarctica, spending only a small portion of their time sucking blood from their hosts. As is the case with all predators, their first priority is to acquire an appropriate target—a cat, dog, bat, bird, rodent, or human. They are not fussy feeders, and unfortunately, their target is often a human. Their target acquisition system includes sensors on their antennae and hairlike bristles that can detect temperature, air currents, vibrations, carbon dioxide, and even female hormones. Several species, in fact, show a strong preference for female hosts. Their aerial attack mechanism is found in their extraordinary jumping capacity.

Fleas fly into the air with a leap so fast that the human eye cannot follow it. These tiny, leg-powered flyers can jump more than 150 times their body length. Usually they execute a nearly vertical leap, although if a target is a little to one side, they can alter their leap to include a small horizontal component. What is amazing is that they can jump hundreds of times per hour and maintain the jumping performance continually for up to three days. Once airborne, they may spin and somersault. If a host is in range, they attach to its hair or feathers by means of the hook-tipped comblike bristles that cover their exoskeleton. Once on the host, their bodies, flattened from side to side, are beautifully adapted to maneuver between the forest of hairs.

Once firmly established, they use their piercing-sucking mouth parts to penetrate the skin; then they inject their saliva, which has both anticoagulant and blood-vessel-dilating components, and activate their suction pumps, siphoning blood into their digestive tube. Some species leave after feeding, while some stay on for further meals. If their flying leap is unsuccessful, they land, usually on their feet, and orient themselves toward the shade. It may be that the host's shadow plays a role in attracting fleas.

Thanks to the lifelong studies of Miriam Rothschild, we now understand how fleas execute their acrobatic leaps, which are so fast that liftoff produces speeds of acceleration of over 100 Gs (100 × gravity). The leap is powered by the elongated rearmost pair of legs. Just prior to blastoff, the flea crouches head down into a cocked position, like a sprinter on the starting blocks. Muscles

alone are incapable of contracting fast enough to produce the kind of acceleration seen in the flea jump. So, how is it done? The answer was found in another
chemical mechanism using elastic rebound. This chemical, reselin, is a protein
that behaves like a superrubber. It can release almost all the energy stored in it
when compressed. As the flea squats in preparation for its jump, the reselin
attached to an arch on the inner aspect of the flea's lightweight elastic external
skeleton is compressed. It is kept in that compressed, energy-loaded state by a
pair of hooked catches attached to muscles. When these muscles respond to a
neural signal to relax, the catches are released and the reselin rebounds, giving
up its stored energy, and then, along with the fast-twitch muscle fibers of the
rear leg, powers liftoff.

Ratborne fleas carrying the deadly plague bacterium have caused catastrophic epidemics among humans since biblical times. Later plague epidemics
ravaged the human population, killing hundreds of millions of people. In the
1300s, the so-called Black Death killed more than 25 percent of the people in
Europe, and in 1665, the Great Plague killed almost 70,000 Londoners. The
plague bacterium is still present in 120 species of rodents today, but humans
have learned to borrow antibacterial chemical weapons from lower organisms.
Antibiotics can at last stop the scourge.

Feathered Aerial Hunters

Recent fossil finds, coupled with aerodynamics and structural analyses, provide
strong evidence that birds evolved from small, terrestrial, high-speed, toothed,
carnivorous, predatory reptiles known as therapods. By comparing the novel
anatomical traits of these ancestors with those of modern birds, it is evident
that therapods (such as the lightweight, bipedal high-speed *Velociraptor*,
immortalized on celluloid in Steven Spielberg's *Jurassic Park*) had already
evolved a subset of features that were needed for flight. Those features were
strong, light, hollow bones; only three working toes; and long, seizing, grasping
arms whose stroke was a precursor to flapping. Although the fossil record is
incomplete, we know that about 150 million years ago a Blue-Jay-sized, birdlike
ancestor appeared, equipped with lengthened arms, modified wrist bones, a
shortened tail, and real feathers. This was a creature named *Archaeopteryx
lithographica*.

Looking at fossils of *Archaeopteryx*, it's hard to imagine how it could fly.
One scenario suggests that this poorly designed flyer climbed to a height and
then, using gravity as a power source, dove off its perch and glided down. The
problems with this scenario are that *Archaeopteryx*'s feet were not adapted for
perching and that there were no tall trees around at that time. A second scenario suggests that bird flight evolved by the animal's running and accelerating

fast enough to create the airflow over its extended feathered forelimbs needed to achieve liftoff. In the course of bird evolution, more changes occurred, producing a flying creature with traits quite different from those of all other vertebrate animals.

Adaptation for Flight

Birds are unique vertebrates equipped with all sorts of weight-saving, gravity-defying structures. They have toothless beaks, hollow bones, perching feet, and wishbones. They have a keeled breast bone that acts as a point of attachment for their powerful flight muscles and stump-like tail bones to which tail feathers can attach. Over the course of evolution, there were other weight-saving changes in the bones of the pelvis, wrist, and hand that made possible a powerful flight stroke and enhanced aerial maneuverability. Perhaps the most critical change was the evolution of feathers, structures whose origins are evident in some of the birds' ground-based reptilian ancestors. For their weight, feathers are stronger than any bony structure and are exceptionally well contoured to act as frictionless lift surfaces. They are waterproof and exceptional insulators that maintain the body heat generated by contraction of the birds' flight muscles. Birds also have the most efficient respiratory system of any vertebrate. This system consists of their lungs and a network of air-conditioning air sacs in their hollow bones.

Given these unique common characteristics, birds diversified and specialized for the survival game. Some became aerial interceptors; some specialized in dive-bombing tactics that took them from the air to the ground and even into the water to pursue underwater prey. Others specialized in low-level attacks on ground targets, ranging from snakes to small mammals. In the process, each adapted to its given mode by developing specialized weapon systems and tactics.

Feeding Strategies and Tactics of Predatory Birds

Ground-Hunting Carnivorous Birds

A number of modern birds capable of flight hunt on the ground for small reptiles, amphibians, and mammals. Two of these are relatively large predators of the East African savannah, the bustard and the handsome, long-legged secretary bird, *Sagittarius serpentarius*. Their favorite prey are snakes, including many venomous

An Aerial Hunter Pursues a Terrestrial Prey
Seeing the eagle's approach, the rabbit runs for cover, abandoning its attempt to escape detection by remaining motionless. However, the swooping eagle, faster than the fleeing rabbit, closes in on its hapless prey, and the rabbit will shortly be impaled within the grasp of the eagle's razor-sharp pointed talons.

species, which are stomped into submission, picked up, battered senseless against the ground, and then swallowed whole. This type of feeding strategy is also practiced by the Roadrunner of cartoon fame, *Geococcyx califoricus,* and the tall, long-legged seriemas, whose eerie, piercing calls ring across the South American grasslands. These seriemas are thought to be descendants of the long-extinct phorusrhacoids, the flightless, huge, carnivorous terror birds that were widespread from 62 million years ago to as recent as 1.4 million years ago.

These terror birds were enormous engines of destruction, the top terrestrial carnivores in the food chain in South America. Some were almost 10 feet tall; had long, powerful legs, which could carry them at speeds of up to 50 miles per hour; and had long, flightless wings which they spread to maintain their balance while in pursuit of their prey. They had long, muscular necks, and the largest of them had a huge, 3-foot-long head equipped with a massive beak, whose razor-sharp edges could open to a gape of almost 20 inches. These awesome eating machines stalked their prey, usually small, horselike mammals, by stealthily moving through tall grasses to cover their approach. Once within range, they would explode from concealment, overtake their fleeing prey, and knock it over with a powerful sideways kick; they would then pick up the stunned prey and batter it to death against the ground. The limp body of the prey was swallowed whole. One of these giant terror birds, *Titanus walleri,* apparently made it to Florida between 2.5 and 1.5 million years ago, but that was the end of the line; all terror birds, small, medium, and large, became extinct. Their role as top carnivores of the grasslands was taken over by smarter, more competitive mammals. (For a more detailed account of the terror birds, see the February 1994 issue of *Scientific American.*)

Feeding behaviors of birds are in part determined by their anatomical equipment, such as specialized bills and talons. But birds also augment their hunting tactics by learning innovative habits and adopting borrowed tools for their armory. For example, Egyptian Vultures, lacking powerful bills, have learned to break open the remarkably strong eggs of ostriches by picking up rocks and hammering on the shell until it breaks. Similarly, greenback herons have learned to use bait to lure fish into their strike range.

No birds have bills strong enough to penetrate the protective armor of shelled mollusks, clams, mussels, and snails. But, a number of gulls have learned to pick up shells and carry the shells high enough to drop and shatter them, thereby making the soft nutritious bodies available. I can attest to their bombing accuracy, having had to sweep dozens of shattered clam shells from a hard-surfaced tennis court every morning. It was certainly better than that of our World War II B-17s, which hit the target less than 5 percent of the time.

Another high-altitude bomber is the Lammergeyer, or Bearded Vulture, of East Africa. These very long winged carrion eaters usually feed after other scavengers have worked over a carcass. They pick up the long bones, soar to a high altitude, and drop the bones on the rocks so that the bones shatter. The Lammergeyers then feed on the exposed bone marrow. Although the Lammergeyers are not raptors and lack talons, they have learned to knock the furry, small rock hyraxes off cliffs, using gravity as a killing weapon.

Birds also learn food palatability. For example, inexperienced Blue Jays that have eaten toxic monarch butterflies and then experienced a bout of vomiting will exhibit one-trial aversion learning. In other words, after once being so affected, they will avoid any moth whose "search image" is like that of the monarch. In this way, other moths whose color and design pattern are similar to that of the monarch are given a measure of protection. However, some Blue Jays and other moth-eating birds have learned to cope with monarchs that contain vomit-evoking doses of toxin by taste discrimination. This skill enables them to separate poisonous from nonpoisonous monarchs.

Some Recent Learned Strategies

Some birds have learned to use other species to help on their hunt. The gaudy East African Carmine Bee-Eater has learned to use the back of the large ground-hunting Kori bustard as a perch. As the bustard moves through the grass, it stirs up a variety of insects. The Bee-Eater launches itself from its flight deck and captures its prey on the wing. Such a strategy saves energy otherwise needed for hunting. The bustard gets nothing in return but does not seem to mind being used as an aircraft carrier. Another group of East African birds called Honey Guides have carried interspecific cooperation to a new level. The Honey Guide locates a bees' nest and then attracts the attention of the ratel, or honey badger, which it leads to the nest. The ratel, with its powerful, long-clawed forelimbs, tears the nest apart to feed on the honey, and as a result, the bird is provided access to the exposed larvae and bees. Recently it has been found that Honey Guides use the same tactic to lead another honey-loving mammal, man, to the bees' nest.

Of all animal species, only humans have learned to make and use fire. But a number of birds have learned to take advantage of fire as an aid in hunting. During the dry season in the East African grassland, there are frequent grass fires—some spontaneous, some started by man. Driven by strong winds, these fires sweep across the savannah, driving insects along the leading edge of the fire into the air. Having learned from previous experience of the sudden availability of food whenever there is a fire, many small, insectivorous birds are attracted to the fire and pick off the insects in flight. As the fire incinerates the dry grass, it leaves behind a denuded landscape and exposes many snakes, lizards, and toads. Some of these animals are singed, some intact, having escaped the heat in furrows and cracks in the soil. However, their cover has been destroyed, making them easy prey for large ground hunters such as storks, bustards, cranes, and secretary birds. These predators also can see the flames from afar and then home in, forming long, advancing skirmish lines to exploit this new and suddenly abundant resource.

Avian Daytime Interceptors

Birds that are visual hunters have eyesight superior to that of all other animals. Their eyes are disproportionately huge, occupying a greater volume in the skull than the brain. Many have eyes that point both forward and sideways. The main

Eagle Eyes   Eagles and other aerial hunters—the raptors—not only have very large eyes, but also specialized optics and retinal anatomy with densely packed bright-light-sensing cone cells. These cone cells not only magnify the target image, but also give it an acuity well beyond that given by the human eye. In bright light, the eagle's visual acuity is at least three times better than ours, and its optics provide it with magnification comparable to that which we get when viewing an object through an eight-power telescope.

optical system, a highly curved cornea and a lens capable of rapid change in shape, focus light rays on a supersensitive retina. The retina, a cup-shaped off-shoot of the developing midbrain, contains specialized light receptors: rods and cones and several layers of nerve cells that process the information provided to them by the receptors. Rods are most effective in dim light, while the three types of cones detect color and provide for an exquisitely sharp color image. Raptorial birds have the highest concentration of cones of any vertebrate. Birds, then, can see details with far greater acuity than humans can. Their long-range vision gives them an image comparable to what a human sees looking through an eight-power telescope. Anatomically, the cones are concentrated in pits called foveas. Humans have one such pit in each eye. Eagles and other raptors have two, one for light coming from the side, giving the bird monocular vision, and the other for light coming from ahead, giving it binocular vision.

In binocular vision, each eye sees a slightly different view or image of the target object. Each eye processes this information in the circuits of nerve cells in the retina. Since birds' eyes have many more layers of data-processing units than the human eye, they can provide more detail to the optic part of the brain. While the eye sees, it is the brain that perceives. Signals from each retina are superimposed in the brain to provide the bird with a three-dimensional color image of the prey. Beyond just image sensation, the retina also has receptors that respond only to movement. But in birds' eyes there is another structure behind the lens, called the pecten, which is comb shaped and throws shadows on the retina in such a way that the slightest motion of a target under scrutiny is perceived and telegraphed to the brain. Because of predatory birds' exceptional capacity to detect motion, many prey species are programmed to escape detection by remaining still, particularly since they can see a flying raptor at long range. Often one can see a raptor playing the waiting game, perched high in a tree, waiting motionless for some undetectable prey to move and give away its position.

Being the swiftest of all hunters, some birds are capable of closing in on their targets at speeds in excess of 100 miles per hour. Such rapid closing speeds impose another requirement on bird vision: it must rapidly adjust the curvature of the elastic lens to keep the target in sharp focus on the retina. This process is called accommodation. It takes the human eye somewhat less than a second to accommodate. But birds can do it in a few thousandths of a second.

Talons and Beaks

Although they are multipurpose, somewhat flexible tools, birds' beaks are also specialized for their specific mode of feeding. Some are suited for spearing fish; others are hooked and ideal for tearing; still others have serrated sharp edges for cutting and holding prey. Eagles, hawks, harriers, kites, falcons, and other

Dive-Bombing Hunters A number of diving birds that hunt in aquatic environments close in on their targets at high speed, and therefore need to actively change the optical power of their eyes (since the approaching targets rapidly appear larger). To do this, they modify the curvature of the lenses of their eyes in a process called accommodation. Diving birds such as the cormorant shown here use their eye muscles to change their lens curvature significantly, providing these birds with useful target images in both the air and the water.

daytime hunters usually hunt small prey that they can carry back to their nest. When foraging, the game plan calls for an energy expenditure less than the energy gained from eating the prey; hence many aerial hunters soar effortlessly when in search of the next meal. Once the target is spotted, the hunter, its wings folded back, starts a sharp, high-speed dive. Those that hunt other birds or bats in the air approach from above and behind. The dive of the peregrine falcon has been timed at over 85 miles per hour. The raptor's killing weapons are the razor-sharp, large talons at the ends of each of its four toes. The prey may be impaled by the powerful grasp of the talons, or it may simply be knocked senseless and fall to the ground, where it is eaten. Other visual raptors hunt ground-dwelling birds, mammals, and fish close to the water. The classic visual hunters using this strategy are the eagles.

Eagles are the largest, most robust diving aerial hunters. Some are medium-sized and narrow-winged, others broad-winged and massive such as the monkey-hunting Crowned Eagles and Martial Eagles of East Africa. The handsome African Fish Eagle and the American Bald Eagle are superb fishers, usually seen perching on tall trees along the water's edge. When they spot a fish near the surface, they launch themselves into a long shallow glide and apply their dive brakes with talons held forward. Once positioned over their prey, they sweep their legs backward, hooking the fish and simultaneously driving their talons into the prey with crushing force. Their wings' tapered primary feathers are capable of fanning out at the tips to act as winglets that increase lift and prevent stalling. Then, with the fish firmly in their grasp, they return to their nest for a leisurely meal.

Some of the other daytime raptors use a different tactic, that of hovering. Kestrels and harriers have been shown to significantly increase their hunting efficiency by hovering and securing a sharp view of the hunting area below them. Once they spot a prey, they drop down in a controlled dive and then use their talons to secure a kill. Hovering, however, uses up lots of energy, and harriers often need three or more kills per day to maintain a positive caloric balance. Three kills a day, week after week, runs up a pretty impressive score when you realize that the human interceptors, namely fighter pilots, only need five kills to reach ace status. Many flying avian interceptors, particularly those that specialize in eating tiny airborne insects, manage five kills in just a few seconds.

Of all avian aerial interceptors, the high-flying, keen-eyed falcons are the most efficient hunters of flying prey. Their standard tactic is to soar at high altitude, conserving their energy until a target comes in view. Their range of vision is over a mile, and once they lock in on their target, they launch themselves in a precisely controlled dive. Wings folded back in order to reduce air resistance, they accelerate, reaching speeds in excess of 100 miles per hour, more than enough speed to overtake their prey. Coming in from above and behind, they sometimes surprise their prey, but at other times the prey spots them and takes evasive action. Small birds can maneuver better than the falcon and frequently escape. Larger prey birds, such as waterfowl, dive down, but escape is often only a temporary reprieve as the falcon climbs and dives again. The bird may be knocked senseless by the impact of the falcon's outsize feet or impaled by the falcon's talons. The coup de grâce is often administered on the ground, as the hunter sinks its powerful notched beak into the prey's neck.

The falcons' hunting efficiency is increased by experience. They seem to spend some time practicing to hone their skills. During the falcon breeding season, males hunt for the females that remain in the nest and spend much more time hunting than out of breeding season. Some falcon species hunt in squadrons along the flyways of migratory birds. Hordes of male falcons waiting near their mates' nests take off just before dawn, flying up and forming a formidable gauntlet, a line with the hunters spaced out over 100 or so yards, waiting for the migrants to come within their airspace.

Individual falcons pick out their prey and dive to the attack, like the Royal Airforce spitfires attacking the Nazi Luftwaffe during the Battle of Britain. During those stressful days, fighter pilots spotting formations of enemy bombers called over their intercoms, "Tally Ho!" to notify their comrades that the target had been sited. There is no evidence, however, that falcons communicate with each other. Massed attacks such as these are highly effective, and even if a prey bird escapes the initial attack, it may quickly come under attack by another falcon. Each male, if successful in its hunt, returns to base (the nest), deposits its kill, and quickly flies off to rejoin the fray. Usually male falcons score a dozen kills a day, a record fighter pilots in World War II never equaled.

The streamlined Swifts and Swiftlets are undeniably the ace of aces of the bird world. Not only are they the fastest fliers, with one species clocked at 110 miles per hour, but they also practice an exceptional maneuver. They begin their patrol by launching themselves from their high perches, often atop tall buildings or chimneys (hence the name chimney swift). Then, they pick up speed and sweep the skies of small insects. They gulp up immense numbers of gnats, flies, and midges and compress them into a ball which is then swallowed. Each food ball can contain as many as 1,000 insects. A single swift can make as many as 20 food balls a day, roughly 20,000 kills per day.

Aerial Hunters of Burrowers

A number of aquatic, terrestrial, and arboreal species of animals have managed to minimize exposure to predators by adapting to a burrowing and tunneling mode of existence. In turn, a number of aerial predators have found ways to exploit these abundant resources by evolving weapon systems to probe into burrows and crevices and extract their prey. Some birds have exploited the burrowing insects and larvae by learning to use long, sharp thorns from cacti as skewers to probe into burrows. The prey are located by both visual and auditory means, and the thorns are held in the beak as the hunting tool. Other birds belonging to the family Picadae, which includes the woodpeckers, have evolved a variety of structural changes for excavating wood-burrowing insects. Their feet have specialized from the usual avian pattern of three forward-facing toes to a pattern of two forward-facing and two rear-facing toes, each toe equipped with a large, sharply pointed claw which is ideally suited for climbing and clinging to a vertical surface. They also evolved modifications of the tail feathers, which serve as a prop when the woodpecker feeds. All birds periodically molt or shed feathers, but the molting of woodpecker tail feathers is staggered over time so that they always have a prop. Their most specialized adaptation is in the evolution of their tongues and beaks.

The woodpecker's primary burrowing tool is its strong, long, pointed beak which it uses to peel away bark and excavate into the wood. Woodpeckers have powerful neck muscles which drive the head back and forth like a jackhammer, producing a rapid and rhythmical staccato drumbeat. Accordingly, there have evolved some changes strengthening this bird's neck vertebrae and cranium, or brain case. In the latter there are also structures that act as shock absorbers so that the woodpecker doesn't knock itself silly when pounding away on a hard wood surface. However, the most remarkable changes are those seen in their extrusible, prehensile, very long, pointed, barbed tongues.

Woodpecker tongues, along with the muscles and bones that support them, are housed within the skull. The bones start just under the nostrils and curl up and over the orbit of the eye, loop around the back of the brain case, and like a loaded spring, curve under the skull and into the base of the lower jaw. The tongue itself is like a long, flexible javelin that has a sharp tip with backward-pointing barbs which secure the prey. In addition, the tongue is coated with a sticky gluelike mucous to prevent the prey's escape.

Hunting in the Dark—Passive Sound Location

Deprived of visual senses in the dark, some aerial hunters evolved a number of specialized night-hunting systems. One of these is an enhanced auditory acuity and the ability to locate sound sources with great accuracy. Sound location requires binaural hearing (using both ears) to locate a sound source within a range of 2 or 3 degrees in azimuth (horizontal direction). Such localization is possible because each ear hears a different intensity of sound and relays this information to the brain's integrating system. The brain then gives motor commands that will bring the hunter in the direction of the sound source.

The ultimate nocturnal acoustic hunters are the owls. These airborne predators can locate the rustling sound created by mice as they scurry along their grass-covered runways, precisely determine both azimuth and elevation,

Listening for the Patter of Small Feet Owls are masters of detecting sounds—even very small ones. Rodents scurrying about in search of food make small sounds, providing the keen-eared owl with target data. Swooping in with wings that muffle its own sound, the owl strikes.

and with incredible speed, swoop in and grab their prey. They can even perform this feat in the winter when the mouse runways are covered with snow. Obviously, such an efficient weapon system required the evolution of some very special equipment.

To detect subtle differences in sound intensity, the barn owl's ears are anatomically located at different distances from the sound source. Since sound waves travel at about 600 miles per hour, it hardly seems conceivable that changing the distance of the ears a few millimeters would make much difference, but it does. The owl's left ear is directed slightly downward, so it is more sensitive to sounds from below, while the right ear is directed slightly upward, so it is more sensitive to sounds from above. The sounds reaching the two ears arrive at slightly different times and with slightly different intensities. Sensory information reaching the owl's on-board computer is processed continuously to produce a three-dimensional map of the location of the moving target. Armed with such detection equipment, the owl can set an intercept course that involves both azimuth and dive angle. However, all instrument systems require some measure of amplification. In the owl, this is provided by specialized facial structures.

Barn owl faces are startling, formed by dense layers of stiff feathers packed tightly together into a heart-shaped structure called a facial ruff. Through the ruff run two troughs, one on each side, extending from the lower jaw to the forehead. Since the densely packed feathers effectively reflect high-frequency sounds, the troughs act as collectors of sound waves over the owl's large facial surface. The sounds are then funneled into the hidden ear openings. Final corrections just before the strike probably use visual information from the owl's strikingly large, forward-facing eyes. Human eyes scan objects by tiny, very rapid eye movements called saccades. In contrast, owls utilize head movements to increase their field of vision.

It is also necessary for night-flying, low-level aerial attackers not to make too much noise on their approach. Owls' soft-edged feathers muffle flight noises so that their approach is silent. They have a formidable weapon system: four curved, ridged, razor-sharp, pointed talons arranged in a square. Thanks to these talons, once a mouse is hit, it remains firmly in the owl's grasp.

The Great Horned Owl, a broad-winged, robust, powerful hunter of dusk, is often seen perusing its feeding grounds from roofs, trees, and even fence posts. Endowed with huge eyes, it can spot its prey, small mammals or birds, at long range. But it is its acute hearing that is its primary target acquisition system. Sweeping along at low altitude, it zeroes in on its target and plunges down with its razor-sharp talons spread and thrust forward. At the moment of impact, it clenches its claws, driving the talons deep into the prey. (Actually, the initial impact is usually of sufficient force to break the prey's back.) With its meal firmly in hand, the raptor returns to its perch to dine at its leisure. Among the small mammals included in the Great Horned Owl's menu are skunks, making this owl one of the few predators undeterred by the skunk's foul-smelling defensive spray.

Ranging the northern forests of North America, Asia, and Europe dwells a rodent-eating specialist, the Great Gray Owl. Slightly smaller than the Great Horned Owl, this owl usually hunts during the day without fear that its shadow may alert its favorite food, the voles. The voles scurry about in their snow-covered runways, feeling quite secure. However, the Great Gray's auditory acuity is

Sound and Sight   In addition to using their exceptionally accurate hearing to locate prey, owls bring their large eyes into play during their final approach toward a target. When homing in on an animal, the owl uses a series of small, precisely controlled movements of its head to acquire both aural and visual data concerning the target's location, distance, and direction of movement.

so sharp that it can detect the sounds of movement under 2 feet of snow from almost 100 feet away. Diving from 24 feet in the air, with its talons held forward but balled up like fists, it punches through the snow crust, creating avalanches that immobilize the rodents in their tunnels. It then plows through the collapsed tunnel, probing until finds its prey. These cold-weather-hunting raptors are amazingly well insulated and have been observed hunting when the temperature drops below −40°F. They prefer to hunt on windless days. Then, the whistle of the wind doesn't produce enough background noise to interfere with noise made by mice under 2 feet of snow. When rodents are abundant, as happens during a population explosion, Great Grays will accumulate in the hunting area. As they deplete the rodent population, they move on to find a new, rich hunting ground, covering as much as 30 miles in one day.

Given the frigid temperatures in the hunting grounds of the Great Grays, one would think that the wings, with their large surface area would cause enormous heat loss. But this doesn't happen. The arteries carrying warm blood to the wings lie next to the veins which carry cooled blood back to the bird's body core. This anatomical arrangement allows for countercurrent heat exchange; this means that the arterial blood transfers heat to the venous blood, thereby preserving body heat without taxing the bird's metabolism.

Echolocating Night-Hunting Aerial Interceptors

The most difficult task for nocturnal airborne predators is to detect, locate, intercept, and successfully engage a moving target. To achieve success, the predator must acquire an almost continuous stream of information giving instantaneous data about the target's size, velocity, direction, and altitude. These data are then integrated with the attacker's speed and direction. Then, an intercept course is determined, allowing the hunter to close in for the kill. The most prodigious night hunters are the insect-eating bats. A single bat can intercept and kill 900 insects in one hour, and as many as 3,000 in one night's mission. Figuring a life span of 15 years, a bat in its lifetime could score an amazing 5 million to 100 million kills.

How do bats achieve this remarkable dogfight success? The bat begins its patrol at dusk, cruising at speeds of up to 40 miles per hour. It ranges the night skies with its echolocating equipment, emitting pulses of sound waves whose frequencies can reach 230,000 cycles per second, more than ten times higher than that of any sound the human ear can detect. While the bat is cruising, pulses are produced at the rate of ten per second by its voice box (or larynx), which has large, ossified cartilages attached to strong muscles. Pull produced by this lever system applies tension to two small vibrating membranes, which in turn produce the sound pulses, each of which lasts only a thousandth of a second. If the sound wave strikes an object, an echo is reflected back and picked up by the bat's extraordinary ears, which are structured to amplify and concentrate the sound. The bat doesn't hear its own sound pulses because a valvelike muscle is synchronized to close the ear during the pulse. The valve then opens, allowing the bat to hear only the echo.

As it starts its night patrol, the bat emits pulses of 10 to 20 per second. The usual range bats use for detecting their targets is about 10 meters. Large echoes indicate a large target, small echoes a small target, and the crispness of the echo indicates the target's texture. If the echo returns at a lower pitch, the bat's Doppler analyzer knows the target is moving away; if the echo is at a higher pitch, the target is approaching. Each ear picks up the echo at a slightly different intensity and time. In this way, the bat's minuscule on-board computer, its 10-milligram brain, automatically triangulates and gives in-flight directions for the intercept.

As the bat closes in on its target, it increases its pulse rate to up to 300 clicks per second; it may even vary the tone (frequency modulation) to get more informational echoes. When the bat is at point-blank range, the target may attempt sharp evasive maneuvers, but the pursuer picks up these changes and makes course corrections at a dizzying speed. The bat is so flexible that it can execute a 90-degree turn in a space slightly greater than its body length.

The bat's weapon system includes the continuous sweep of tail and wing membranes, which allows it to make a kill when it comes within a given proximity of its target. The bat then scoops the insect in its wing or tail webs and stuffs it into its mouth. During this time, some species of bats emit pulses through their open mouths; others stop sending hunting pulses. Still other species of bats send pulses through their noses, enabling them to acquire new targets while devouring their earlier kill.

From first echo to kill, the entire hunt takes only 3 to 4 seconds. A bat's sonar power-to-weight ratio makes it about 1,000,000,000,000 times more efficient than our best sonar detector systems. It can even pick its target out of a number of decoys. Donald Griffin, the discoverer of bat echolocation, showed that with a little training, bats could pick out of the air a mealworm that had been thrown up along with a handful of pebbles. How our antiballistic missile engineers would like to match that feat in picking out real warheads from decoys!

How efficient is the bat as an interceptor? Even those insects that can detect bat sonar and take evasive action are hit 60 percent of the time. Insects without sonar detectors fall victim 90 percent of the time. During their night patrol, bats feed almost continuously. Their digestive tracts require only 15 minutes to process each insect.

Since flying insects don't fight back, it would hardly be fair to call their air-to-air combat a dogfight. However, there are some animals that do hunt insectivorous bats while they are in flight. One of their most deadly enemies is the

bat-eating bat, a large, ferocious echolocator. Another bat-eating aerial interceptor is the powerful, bat-eating cave hawk, a visual predator that hunts at dusk.

Surface-Hunting Bats

How do you hit a target at low level in a maze of obstacles and in difficult terrain? It's no problem for the frog-eating bats of the South and Central American rain forests. They, too, use echolocation in finding prey and can avoid colli-
sion by obtaining a detailed acoustic image of their surroundings as well as
their target.

These ground-hunting bats have an incredibly accurate discriminator
which allows them to navigate even in a space full of rods and wires. Laboratory experiments tested bats in a darkened room with a maze of wires less than
2/10 millimeter in diameter. Not only could they evade the wires, but they
could tell if the wires were vertical or horizontal, although they did much better evading horizontal obstacles than vertical ones. What is amazing is that they
could adjust the timing of their wing beats to be just at the right angle to allow
them to fly through a narrow opening. Ground-hunting bats have been
observed flying right through tousled thickets to snatch their prey.

The false vampire bat hunts ground-dwelling rodents in such a tangled
maze, and the fringe-lipped frog-eating bat uses similar skills to avoid obstacles. For example, some frogs dwelling on the tropical rain forest floor are the
highly toxic, colorful dendrobatid tree frogs. Bats, detecting these frogs' calls,
know to avoid this kind of lethal meal. They are also able to detect the toxic
frogs with the sensory receptors on their lips and abort the attack at the very
last moment.

The fish-eating bat, *Noctilo,* is a bizarre-looking night hunter with an
extraordinarily ugly, bright orange, hairy face, large ears, and legs equipped
with long, sharp toe claws that can be rotated to face forward. At night these
graceful flyers can be seen flapping just above the surface of the placid waters
of South American rivers and ponds. As they fly, their mouths open to emit
pulses of ultrasonic sound waves. Fish swimming close to the surface create
ripples which the sound pulses bounce off of, and their echoes are detected by
the fish-eating bat's enormous sound-amplifying ears.

Once they have detected their target, the bats increase their rate of pulsing
to get more information about the speed and direction of their prey. Swooping
to surface-skimming heights, they lower their rear legs with their curved, forward-pointing, needle-sharp claws and gaff their prey, which they promptly
stuff into their toothed mouths. After capturing their prey, they return to the
roost to dine at a leisurely pace.

Blind Fishing   Some species of bat possess echolocating equipment that is sensitive enough to detect echoes reflected from ripples in the water's surface. Such ripples are usually created by fish swimming near the water's surface. The bat homes in, lowers its large, clawed rear feet, and gaffs the fish.

Countermeasures and Evasion

As bats evolved more efficient hunting systems, their prey, under selection pressure, invariably developed systems to counter them. A perfect example of this is seen in the ability of some moths and other insects to detect the ultrasonic hunting pulses put out by marauding bats to locate their targets. We have observed bizarre flight patterns of night-flying insects for many years. It appeared that these maneuvers were somehow associated with attacks by night-hunting bats. If these apparent evasion tactics were triggered by detecting the bats' sonar, then some sort of sensor must have been capable of detecting these ultrasonic pulses (8,000 to 230,000 cycles per second).

Dr. Kenneth Roeder of Tufts University was able to show that the ears of a number of species of moths are capable of tuning in on the ultrasonic frequencies used by most bat sonars. The moths use this information to determine both the direction of the bat's flight path and its proximity. As a matter of fact, these moths can pick up bat sonar at distances (40 meters) beyond the bat's echolocating range (1 to 10 meters). In such cases, the moth simply turns tail and flies away from the bat, evading detection. However, if the moth's acoustic early warning system picks up bat sonar that indicates the interceptor is closing in fast, the response is more complex.

Since bats fly faster than the maximal speed of moths, to flee in a straight line would be disastrous. The insect's usual response is to fold its wings and dive for cover as the bat locks on to its target. However, if the insect's behavioral repertoire were limited to only this response, the bat could learn the moth's response and easily make corrections. There would be an obvious survival advantage for those insects whose evasive responses were unpredictable. Consequently, sonar-detecting insects evolved a spectrum of evasive maneuvers.

In addition to an early warning system and evasive maneuvers, some moths may be able to jam the bat's sonar by emitting volleys of ultrasonic clicks. These sonar countermeasures have been shown to cause some hunting bats to abort their attacks. It is not known whether the attacks are halted due to the insect's jamming the bat's sonar, to a startle response on the part of the bat, or to a learned response that the target tastes bad. A number of moths, such as the dogbane tiger moth, are known to contain foul-tasting, irritating chemicals; perhaps the clicks emitted by these moths simply warn hunting bats of their disagreeable taste.

In examining the capabilities of bat sonar, we see an extremely versatile system capable of varying the pulse rate, pulse frequency (8 to 230 kilohertz), pulse duration, and pulse intensity (up to 100 decibels at 10 centimeters in front of the bat). By emitting different calls in certain conditions, the bat can

Listening to the Dark   Night-hunting bats are 90 percent efficient in intercepting and killing flying insects. However, for every weapon system, countermeasures have evolved. Thus, many species of insects possess auditory early-warning systems with greater ranges than that of the bats' echo detectors. When predators approach, such insects either fly away or carry out rapid evasive maneuvers. Insects with these systems fall prey to bats only 60 percent of the time, clearly an improvement in their odds of survival.

increase the amount of information it is receiving. Using frequency modulation provides the bat with precise target resolution. Also, by utilizing a Doppler-shift compensating system, the bat can determine the velocity of its target.

The increased sophistication of the bat's sonar was probably an evolutionary response to insect countermeasures. Indeed, some bats counter insect early warning systems, becoming less acoustically conspicuous by increasing the frequencies of their hunting calls above or below the narrow bandwidth that moths can detect. The history of the insect-bat weapons race and the countermeasures employed are remarkably similar to those of human aerial combat.

In some insect species, there is a bizarre evolutionary adaptation involving the detection of bat sonar; these insects harbor a parasite that destroys only one ear. Apparently, the parasite follows some sort of chemical signal for only one ear. It is presumed that initially this parasite attacked both ears, rendering the insect highly susceptible to predation. Those insects with only one ear destroyed, however, could still detect the bat sonar. Because of natural selection, those insects were able to pass on their genes. The one-sided parasitic targeting system, then, is an extraordinary example of adaptation to predation pressures.

Orientation and Aerial Navigation

One corollary of flight for both predators and prey is the need to leave their nests, forage over distance, and then find their way back. Getting there and back requires the ability to form spatial memories. Birds are particularly good at learning to use visual landmarks and landscapes for both local travel and long-range navigation during migrations. They can remember rivers, lakes, and coastlines; this enables them to follow well-defined corridors or flyways during migratory flights that cover thousands of miles. Spatial memories are clearly learned; inexperienced birds get lost frequently. Spatial learning appears to be dependent on brief experiences during some critical period in the bird's early life. Such early memories are imprinted in the bird's brain in the form of firmly established long-term memory circuits. This imprinting can occur within 40 hours of hatching and is also related to establishing specific relationships later in life. So, if the first moving object a newly hatched duckling sees is its mother, it imprints, "I am a duck; I am a mallard."

The discovery of imprinting and its effect on subsequent behavior won the Nobel Prize for Konrad Lorenz. Sharing the Nobel Prize with Lorenz was Karl von Frisch, who showed that bees could use the Sun as a guidepost from hive to food source and back.

Soon after this discovery, it became apparent that birds could navigate over even very long distances devoid of landmarks. Researchers suspected that they used the Sun as a guidepost, because studies on homing pigeons showed that they became disoriented on overcast days, when they couldn't get a bearing on the Sun. Subsequently it was shown that birds not only derived directional guidance from the Sun, but they could adjust to changing solar positions at different times of day. So, the Sun compass works because birds have a built-in clock, a time sense that compensates for the Sun's 15-degree-per-hour move-

Go Three Rays Down and Take a Left   Honeybee scouts are incredible navigators and they give excellent directions as well. When away from their hives they use their color vision and olfactory sense to locate food sources, and, relying on memory, use polarized sunlight to find their way home. Individual scouts then recruit large numbers of hivemates and give them— by performing a "waggle dance"—instructions concerning the direction (and distance from the hive) of a given food source, as well as the relationship between the position of the Sun and those of the food source and the hive. Recruits learn and remember these instructions, outlined not only by a scout's motor activity but also by its acoustic signals.

ment across the daytime sky. Solar navigation has been convincingly demonstrated not only in flying birds but also in penguins in the landmark-free Antarctic ice cap. When taken from their breeding grounds and flown to the interior of the ice cap, the penguins headed unerringly back to their nesting position, as long as they could see the Sun, that is. On cloudy, sunless days, they became disoriented and wandered aimlessly.

However, there are night-flying birds that show remarkable feats of navigation in the dark. At first, researchers were stymied by this ability. What was the source of their directional information? Could it be they were using the stars as celestial guideposts? Subsequent experiments showed conclusively that some birds can indeed use celestial navigation, not only in the northern skies but also in the southern skies, whose stellar anatomy is quite different. In the Northern

Hemisphere it has been shown birds use an array of constellations within 35 degrees of the fixed North Star, Polaris: Draco, Cassiopeia's chair, and the Big Dipper.

I remember learning these constellations and their relationship to the North Star when I was an aerial navigator cadet in World War II, and now, more than 50 years later, I can still recognize them. That a tiny bird could also learn the constellations is mind boggling. Why not just use the North Star? The redundancy is useful on partially overcast nights that might obscure Polaris, when you can still determine which way is north from the other constellations, which, although changing their positions continuously, maintain a constant relationship to the North Star. Apparently, different species of birds use different constellations, on the basis largely of previous experience.

In totally overcast skies that obscure a view of the stars and in darkness that prevents the use of landmarks as guideposts, there are birds that still make their way toward their destinations with considerable accuracy. Could these birds be using some sort of built-in magnetic compass or some sort of map sense in which they know where they are in relationship to their goal? A number of laboratory studies have shown that several species of birds can indeed navigate by sensing Earth's magnetic fields. This geomagnetic detector system, while not understood at a cellular level, is extremely sensitive to a magnetic pull of as low as a hundred-millionth gauss (a measure of magnetic field strength). However, we have yet to demonstrate that birds have the kind of map sense that humans have.

Birds are not the only fliers capable of long-range aerial navigation. Bees, too, are known to use solar navigation. Using their compound eyes, they establish the relative spatial relationships between their home base, the hive, a distant food, and the position of the Sun. Furthermore, on returning to their hive, bees communicate this information to other foragers by means of a series of movements referred to by Von Frisch as the *schwanzeltanz,* or tail dance. Demonstrating this remarkable capacity to use solar navigation and communicate spatial information won Von Frisch the Nobel Prize. Beyond the birds and bees, there is one more group of long-range migratory flyers most certainly worthy of mention: moths and butterflies.

About fifty species of butterflies and moths make periodic migrations. The European painted lady flies back and forth across the Mediterranean from the Sahara Desert to the British Isles. The monarch butterfly of North America migrates annually from Central America to New England and back. How they do this is a mystery, but it has been suggested that invertebrate animals with compound eyes can look at an overcast sky and still determine where the sun is from the way sunlight is polarized by passing through a somewhat opaque atmosphere. A recent study reported in *Nature* that the monarchs do indeed use a sun compass.

The Scent of a Gypsy   Many female animals emit distinctive mixes of molecules into the air to call males of the same species to home in and mate. The males' chemical sensing capacities are often astounding and sometimes they can detect females hundreds of meters downwind. For example, while the air may be filled with the odor messages of many other species, a male gypsy moth's chemoreceptors—tens of thousands of them, located on its huge antennae—allow the moth to home in only on females of its own species. When in flight, male gypsy moths also use visual references as part of their downwind navigation system and it is even possible that they use ultrasonic acoustic signals at close range. The gypsy moth's navigational system for finding mates is so efficient that the gypsy moth population experiences periodic explosions, during which hundreds of millions of their offspring denude the trees.

During World War II, I was an aerial navigator on long-range bombers. Our flights took us over landscape-free oceans and deserts. We were well trained and had accurate maps, precise chronometers, compasses, and celestial tables that permitted us to use both solar and celestial navigation. We also had long-range radio aids to navigation and radar which worked sometimes. I was pretty good as a navigator, as attested to by the fact that I survived and eventually got my plane home, but in all honesty, there were many times when I was totally lost. It is humbling to realize that birds with pea-sized brains and some flying insects with brains the size of a pinhead can perform navigational feats that surpass that of experienced World War II navigators. Today, utilizing vast improvements in instrumentation, we humans can navigate as well if not better than birds, but we surely can't match them for the miniaturization of their navigational instruments.

Another problem involved in long-range flight is to get to your objective without running out of fuel. Literally thousands of members of flight crews died because of errors in navigation or erroneous calculations of fuel consumption. Birds and insects can apparently sustain long-range flights by making physiological adjustments that fine-tune their available fuel supply, loading up with high-calorie fat before departure. Monarch butterflies are virtual butterballs, with fat's making up about a third of their weight.

Flying Reptiles—The Pterosaurs

According to the best and most recent paleontologic evidence, the dinosaurs, pterosaurs (flying reptiles), and modern birds all evolved from a small, two-legged, lightweight ancestral reptile that inhabited the earth about 250 million years ago. How the pterosaurs evolved to fully qualified aerialists is still a matter of speculation, but evolve they did. Fossil records of small, bird-sized, toothy, long-tailed pterosaurs date back some 225 million years. Their airframes were lightweight and their wings were based on an astounding evolution of the fifth finger of the hand which was longer than their entire body. This elongated bone supported a membranous wing. The other fingers were normal in proportion, claw-tipped and capable of grasping; some pterosaurs may have used them to hang upside down from a perch, as modern-day bats do. These early flying reptiles also had long, thin tails that served to stabilize their flight. Judging from their toothsome heads, they were most probably fish eaters, and some of the later forms may have been warm blooded (although this is purely speculative). Pterosaurs ruled the skies for about 50 million years. Then, according to the fossil record, about 180 million years ago, a new, more aerodynamic body plan emerged, in the form of the wing-fingered pterodactyls.

When Reptiles Ruled the Skies (facing page)   Long-tailed flying reptiles called pterosaurs appeared about 225 million years ago and ruled the ancient skies until they were replaced by the short-tailed and much larger pterodactyls (shown here). One pterodactyl, *Quetzalcoatalus*, weighed 150 pounds and had a wingspan of 40 feet. Pterodactyls became extinct approximately 65 million years ago.

Pterodactyls had long, flexible necks, much longer heads, and lighter skeletons than their predecessors. The bigger heads holding bigger brains provided the fine motor control necessary to stablize flight. As a result, the stabilizing tail structures of their ancestors became expendable, as did the early pterosaurs that, for unknown reasons, became extinct about 140 million years ago.

Now the unchallenged dominant species, the pterodactyls exhibited a massive population explosion. They also diversified, creating new, bizarre forms, some of which were enormous: *Pteranodon sternbergi* had a wingspan of 25 feet, *Cearadactylus* had an 18-foot span and an enormous, toothsome, mouth, and the giant of all flying creatures, *Quetzalcoatalus,* weighed 150 pounds and had a wingspan of 40 feet. In all probability, the pterodactyls occupied the same ecological niches as the modern birds. Their large heads showed all kinds of specialized feeding structures. Some think *Quetzalcoatalus* was a wading predator that hunted in shallow waters like modern-day herons and egrets. Some may have been skimmers, some aerial hunters, and some carrion eaters. Certainly, most were exceptional soarers whose long, thin wings made them aerodynamically highly efficient. With their long wings, however, it would be unlikely for them to be diving hunters like terns or pelicans, nor would they do well in the tangle of a forest. Some of the larger pterodactyls had large head crests that they possibly used as flight stabilizers or air coolers. More likely, they used these as sexual display devices to entice females of similar inclination.

Coexisting with the later forms of flying reptiles and probably competing with them, were the true birds with feathers. Suddenly, about 65 million years ago, at a point in time called the Cretaceous-Tertiary boundary, a cataclysmic extraterrestrial impact occurred in the region of the Yucatan. The subsequent results of this impact led to the extinction of the pterodactyls, leaving the insects, birds, and bats to rule the skies.

Countermeasures against Aerial Predators

Defense against aerial predators begins with detection before the hunter is within attack range. Such early warning systems often involve a type of learning at a critical period early in life. Young birds become motionless when they see any aerial object such as a falling leaf, a pigeon, a duck, a goose, or a hawk. Some early experimental studies suggested that they could discriminate between the silhouettes of raptors and nonhunting birds. Researchers at first concluded that this type of learning was similar to that of humans who learned to identify friendly and enemy aircraft silhouettes flashed on a screen for only a tenth of a second. But that gave birds too much intellectual credit. What actually happens is they learn to ignore familiar objects, undergoing a process called habituation, in which the nervous system tunes out monotonous, repetitive stimuli. Raptors, however, are uncommon and thus continue to evoke the crouch-and-freeze response. Since sharp-eyed aerial hunters' eyes are highly sensitive to motion, it's not surprising that many prey have evolved a built-in freeze response.

Ground-dwelling and perching birds are always scanning the skies with their own acute vision. I was watching a covey of quail feeding. Suddenly, they all stopped and appeared to scan the sky. I looked up, and there, high in the air, was a bird. Training my binoculars on it, I saw that it appeared to be a hawk. Social animals, flocks of birds, troops of baboons, colonies of burrowing prairie dogs, and meerkats all seem to employ sentries. While studying prairie dogs in South Dakota, I was struck by the amount of time they spent standing up and scanning the sky. If a hawk appeared, the number of sentries on the alert increased. Should the raptor show any sign of attacking, the sentries would sound an alarm and all the prairie dogs would dive into their burrows. Similar behaviors are seen in the South African meerkats that live in small troops in open semidesert country. These small, sharp-nosed mongooses are day hunters. Laboring in such exposed terrain, they are highly vulnerable to aerial attack. In order to counter this threat, there is always at least one meerkat on guard duty, standing erect on its hind legs often on the top of a tree and scanning the sky for hawks, eagles, and kites. The meerkat doesn't sound the alarm if a familiar nonpredator flies over; but if he spots a raptor in attack mode, the alarm is sounded and the colony takes cover. In this case, the ability to distinguish friend from foe depends on habituation, so a novel airborne object such as a low-flying airplane or helicopter will trigger alarm-escape behavior.

I've witnessed another early warning response to aerial hunters in fiddler crabs, common to mud flats in New England. They live in burrows but forage in the open, where they are vulnerable to seagulls. The crabs have two compound eyes at the end of long eyestalks; with these, they can scan the whole sky. I noticed that their usual scurry-about style of foraging would immediately stop when a shadow fell over them. I recorded the fiddler's eye response electronically and found that a burst of electric activity appeared when a light was turned on; but when the light was turned off, there was another burst of electric activity, an "off" response such as would be evoked if a shadow fell on a brightly illuminated crab. The sensory receptors' "off" response in turn produced a freeze-in-place signal, making the crabs less vulnerable to the predator's motion detectors. Such responses appear to be innate, rather than due to habituation learning. However, some animals actively learn to identify friends and foe.

Baboons in East Africa spend most of the daylight hours feeding on the ground and most of the night hours in the relative safety of trees. They are preyed on by a variety of terrestrial and aerial hunters, such as leopards, pythons, and several species of monkey-eating eagles. At all times, some members of the baboon troop, including juveniles, are on the alert for potential predators. Each type of predator evokes a different kind of alarm call and response. If the predator is a monkey-eating eagle, the alarm causes the troop to flee to the lower branches of the tree, where they are not likely to be attacked. It appears that the process of learning to recognize the type of predator and the appropriate alarm is taught by older, experienced members to juveniles, and if in learning the juveniles make a mistake, they are scolded. While the strategy of recognition of threat and escape is a sound strategy, there are other more active deterrent strategies. The famous French general Marshall Ferdinand Foch once said, "The best defense is a good offense." This is a strategy adopted by animals long before humans set foot on this planet.

Deception

In all arenas of predator-prey interactions, the evolution of deception strategies is common. Some of these tactics call for animals to make themselves less conspicuous. In the case of ambush-style hunters, their ability to escape detection by prey confers on them greater hunting efficiency. In the case of prey animals, their ability to be invisible to visual predators increases their chances of survival. So, it is not surprising that camouflage, a combination of inherited alterations in both shape and color, produced organisms with an uncanny similarity to their physical background.

Genes control both color and shape, and visual hunters, many endowed with color vision, learn to recognize colors, patterns, and shape in order to create target images of their preferred food. Prey adapt both by color change and by selecting backgrounds that match their colors. These adaptations can evolve with amazing rapidity. For example, nocturnal peppered moths, with their creamy speckled wings, rest during the day on lichen-covered tree trunks, where they are beautifully camouflaged and safe from avian predators. Around the turn of the last century in industrial areas of England, the peppered moth population had suddenly changed to a color pattern of black speckled wings. Why did this happen? Heavy industry produces a multitude of pollutants, including sulfuric acid, which kills lichens. Denuded of the white-speckled lichens, tree trunks appear dark brown or blackish due to soot deposits. Under these conditions, which existed in England at that time, creamy-winged pepper moths became prime targets for birds and were preyed on incessantly; however, their undetectable black-winged mutants, well-served by protective coloration, increased in population.

Most colors of vegetation are varying shades of greens and brown, so it is not surprising that many animal species evolved green or brown camouflage. Over the course of time, a great many organisms adapted with extraordinary refinements in color change. The skin of many insects, frogs, toads, and lizards contain pigment cells, within which are colored granules. The most common of the pigments is a brownish-black substance called melanin. The pigment can either be concentrated into a tiny speck or be dispersed into intricately branched projections of the cell, allowing the animal to change color to match the background with incredible rapidity. Most of these changes are initiated by the eyes. Detecting a color change, the eyes trigger neural and hormonal changes that either disperse or contract the pigment. A chameleon lizard or a praying mantis can respond to a new background almost instantaneously, harmonizing with its surroundings as it waits in ambush. But the world is not all green and brown, and in other environments, the evolution of camouflage reaches bizarre extremes.

Some species of insects look like twigs, some like leaves, some like flowers, and some even like inedible bird droppings. Others camouflage themselves with debris. Yet another effective strategy is to become inconspicuous by countershading with light underparts and dark upperparts. However, many backgrounds are not monochromatic but are textured and exhibit many shades, textures, and shadows. So, adaptations involve spots, stripes, and speckling. Much of the world is also snow covered for some, if not all, of the year. In the cold environment it is the warm-blooded animals, birds, and mammals that adapt to winter white to blend with their environment. But, what causes seasonal color changes? In some cases, it has been shown that seasonal clocks, whose set point is adjusted by day length (the relative number of hours of daylight per day) to activate neural and hormonal changes, can cause birds such as the winter grouse to molt. When day lengths are long, for instance, white-adapted winter grouse lose their white feathers and replace them with brown, but when day lengths shorten, they molt again and replace the brown with white. Such reactions are photochemical and not affected by temperature.

Advertising Toxicity

Many animals and plants evolved chemical weapon systems that discourage predators either by being unpalatable and foul-tasting or by releasing chemicals that make the predator sick or kill it. Possessing such potent chemical defenses is only useful if it deters the predator's attack. So, they evolved mechanisms—visual, auditory, or olfactory—that advertise their toxicity. Some of the most beautiful and colorful organisms in nature possess deadly toxins.

Many poisonous animals make no attempt at concealment. Their protective strategy is the use of bold colors—oranges, reds, and yellows—and prominent black stripes, bands, and spots as warning patterns. This is not to say that conspicuous patterns and colors are always a warning of toxicity, since many species rely on the same strategy to advertise their sexuality when seeking mates. Predators learn by painful experience which colorful prey are dangerous.

Deceptive Love   The mating procedure of the various species of fireflies (actually flying beetles) involves a series of light-flash signals. A firefly's light-producing organs are under its brain's control, and males flying the night skies flash in specific sequences to notify the ground-dwelling females of their amorous desires. Females send back a coded sequence of flashes which signal acceptance as well as an invitation. However, the females of some species have learned to imitate the flash sequences of other species, and use them to entice unsuspecting males to "come and play"—but when these males land, they are instead eaten!

Bees, wasps, and hornets advertise with bold yellow and black, or white and black stripes that they are armed and dangerous, equipped with potent defensive weapons. Even though the working bee dies in the process of stinging, it still helps to ensure that the nest bearing the queen is protected.

Behavioral Countermeasures against Predators

One of the most successful strategies used to deter predators is to go on the offensive and attack the would-be predator. Such attacks do not necessarily inflict serious damage on the predator, who is usually bigger and stronger than the attacker. Instead, the goal of the attack strategy is to annoy, harass, and otherwise distract and discourage the predator. Such strategies are common in birds needing to protect their vulnerable nestlings and eggs from a great variety of hunters that would exploit the nutrient-rich resources of the nest. Many birds have a built-in, hardwired attack program that is activated when predators approach their nest. Female turkeys, for example, are programmed to attack any intruder that moves in the vicinity of their nest. A predator faced with an aggressive mother turkey usually avoids attack and retreats. The only thing that turns off or vetoes the mother's attack program is the sound made by turkey nestlings. If as an experiment, you block the ears of a female turkey, she will even attack and kill her own young.

Many birds, both male and female, practice attack behavior against nest robbers. I have been harassed and screamed at by a variety of shore birds when I have inadvertently walked into a nesting area on the beach. Blue Jays are particularly noisy, persistent, and aggressive in defense of their nests. Even large, fearsome predators such as hawks and Great Horned Owls are repeatedly mobbed by smaller, highly maneuverable birds. A perching owl seems defenseless against this constant harassment and almost always flies away in defeat. However, many fast-flying small birds will continue to attack the retreating predator. Sometimes singly, sometimes in groups, they repeatedly attack from above and behind, badgering the slower, less maneuverable predator. I've even seen Red-Winged Blackbirds actually land on the back of a fleeing crow and peck him a couple of times before taking off to execute another attack. Harassment as a tactic is also practiced by scavenger birds such as vultures and Maribou Storks.

Anyone who has spent any time in the East African savannah has seen the carrion eaters swarming to the site of a predator's kill. They form a circle around the feeding predator, constantly harassing him as he is feeding. Often, the predator, his patience worn down, loses his composure and tries to attack the annoying pests, but they are alert and adept at avoiding such assaults. Cheetahs are particularly sensitive to such tactics and, when faced with a horde of vultures, will quickly abandon their kill. However, the most effective form of harassing behavior is practiced by the small songbird, the Fieldfare.

Crows and their kin are adept at finding even well-camouflaged songbird nests; they make their living on eggs or helpless nestlings. A Raven approaching a Fieldfare's nest is spotted by the alert female; her alarm call recruits her nearby mate. The two birds then launch themselves from their perches, dive toward the Raven, and release a blob of foul, sticky feces. The accuracy of these dive-bombing runs is truly amazing, far better than that achieved by dive-bombers in World War II. Somehow, these tiny songbirds can calculate the velocity and trajectory of their excretory missiles and time the contraction of their rectal muscles with such precision that they hit their target more than 50 percent of the time. On being so attacked, the bewildered Raven remains motionless, transfixed by the repeated splattering, and, mouth agape, screams his defiance. That is a big mistake, because the bombing accuracy of the birds is such that sometimes the blob of feces scores a direct hit down the hatch. Eventually the gagging Raven, realizing that the price of this meal is too high, beats a retreat.

Another behavioral tactic practiced by a number of birds involves distracting the predator and luring it away from the nest that holds its vulnerable young. Rudyard Kipling, in his story of Rikki-Tikki Tavi, describes how Darzee, the Tailor Bird, attracted a hunting cobra's attention and then, feigning injury, led the cobra away from the nest. When the cobra closed in for the kill, the apparently wounded bird suddenly recovered and flew off. A number of animals practice such a tactic, and it usually works.

Deception theatrics are also practiced by a number of prey animals that feign death when approached by a predator. This tactic is successful because a number of predators only attack living things and only eat what they themselves have killed. Such a behavioral ploy is called "playing possum" because the American marsupial, *Didelphis virginianum,* commonly known as the opossum, when attacked, will become immobile and appear to be dead. But, the Academy Award for best actor at playing dead goes to some species of the common garden snake that not only go limp and motionless when threatened but also assume the look of death, complete with mouth agape, tongue hanging out, and eyes fixed in a motionless stare. Once the predator departs, resurrection occurs and the snake lives another life.

Deadly Eggs

A number of predatory insects of the orders Diptera (flies) and Hymenoptera use a novel weapon system, a munition that slowly kills the prey from within— their eggs. These animals are called parasitoids, and they prey only on very specific hosts. One such predator is the large, solitary wasp, called the tarantula hawk, that stalks the large, deadly tarantula spiders. Both adversaries are armed with highly poisonous venoms, the tarantula in its fangs, the wasp in its stinger. The battle is fought on the ground, and the victor is usually the wasp. However, the wasp's venom doesn't kill the spider but paralyzes it. Then the wasp, an excellent digger, excavates a burrow, drags its paralyzed prey into it, and lays a single egg on the spider. After closing the entrance to the burrow, the wasp takes off. The developing wasp larva then feeds on the food supply left by its mother. Another wasp uses the same strategy, but its prey are specific species of cicada.

Another species of wasps is called ichneuman flies. These narrow-waisted flyers carry a long, thin tube at the end of the abdomen, which serves as both a drill and an egg-laying tube. This wasp locates its prey—another insect, insect eggs, or insect larva—drills into the host, and injects its eggs. As the egg develops into ichneuman larva, it eats the host from within and almost always kills the host. The prey are highly specific and include host larvae that develop in wood tunnels. Ichneuman flies locate their hidden prey by first detecting a distinct odor produced by a fungus associated with the wood-boring larvae. Then, using the saw-toothed tip of the ovipositor, they drill through the wood (a process that may take as much as a half an hour) and insert an egg into or near the host larvae. The ichneuman larva then consumes the host larvae. Those ichneuman flies which prey on wood-boring insects have impressive ovipositors that are seven times longer than the wasp. What is remarkable is the extreme precision of prey location provided by this unusual weapon system.

Among the Diptera that are parasitoids are a number of flies that attack a variety of prey, including mammals. The infamous screwworm fly burrows into cattle skin to lay its eggs. Its developing larvae (maggots) cause the loss of millions of dollars at cattle ranches. Another parasitoid fly specializes in attacking snails. In the next chapter on parasitism, we consider a number of parasitoid flies whose larvae develop in humans.

Chapter 6

Eating from the Table of Others

The usual scenario for the survival game played out between predators and prey is one in which the predator, if successful, kills the prey, and the payoff is a supply of nutrients for the hunter. However, this is not the only game plan. Indeed, in the first billion or so years of life on this planet, there were no true predators, but there were other survival strategies that evolved. One strategy involved forming colonies of high relatedness in which cooperation provided for improved chances of survival, as the old adage says, "There is strength in numbers." Cooperative strategies by like cells begun by the early ancestral life-forms were subsequently utilized by more advanced, multicellular organisms. But another way in which this cooperative tactic evolved was that two different species of organisms formed a mutually beneficial interaction. This kind of mutualism provided a relationship that was symbiotic; that is, each organism provided something and received a payoff in return, thereby providing benefits for both. This strategy proved to be a highly viable game plan, more robust than that practiced by individuals acting in their own self-interest.

Clearly, the ability to form groups of individuals and the ability to establish symbiotic relationships both required an ability to recognize friend and foe. Identification relies on some sort of memory process, but the early organisms lacked nervous systems that could store memories. Instead, their memories were based on the recognition of specific molecular signatures on their surfaces. Encounters between like cells or cells of symbionts were random, but if such an encounter did occur, recognition and bonding were immediate. Certainly, in such random meetings, prolific reproduction was a prerequisite for increasing the improbable odds of interaction. Other biological factors affecting the probability of a successful encounter would obviously have to include relative mobility and longevity of the interactants.

The early bacteria certainly had motility. They could glide over surfaces or swim by the movement of clusters of whiplike structures called flagella. These flagella were structurally quite different from those found in more-advanced, true cells, but they did the job. Bacteria could move toward a chemically attractive site or move away from a potentially noxious site. So, one could say that the universal strategy of fight-or-flight behavior began early in the history of life.

Competition, a driving force of evolutionary processes, exists between members of the same species not only for the resources necessary for survival but also for mates that will carry on an individual's genetic program. In bacteria, sexual reproduction does not exist, although there is some exchange of genetic material between individuals of some species. So, sexual selection does not take on the competitive nature of that seen in organisms that reproduce sexually. However, competition between species for the same resource, such as food or space, has existed for 3 billion years. One of the favored strategies of a species was to develop a chemical weapon that was harmless to members of its own species but was destructive to other species near it. Releasing such toxins or poisons into the environment may have been constant or may have been activated by the perception or, in this case, presence of a competitor. Humans later learned to exploit the ability of bacteria, particularly those of the genus *Streptomyces,* to produce antibacterial toxins known as antibiotics to supplement the human body's already awesome defense mechanisms for defeating disease-causing bacteria. Chemical warfare in a vast variety of forms evolved in later organisms, particularly in plants and fungi, which lack the ability to escape plant-munching herbivores.

Later, after symbiotic relationships evolved and genes for cooperation between species existed, a number of mutations occurred that produced renegade species of symbionts. These renegade species defied the rules of reciprocity, in that they decided not to give back anything to their host. They took but did not give; so, they were no longer symbionts but were parasites. Parasitism as a strategy had several requirements: First, there had to be some sort of mechanism that allowed the parasite to recognize a specific host organism. Most parasites are highly specific for a given host. Second, parasites had to be able to find their hosts. (Viruses, the first parasites that parasitized bacteria, relied on random meetings of their hosts; such meetings were highly improbable events.) Third, in order to improve the odds of finding a suitable host, parasite fitness depended on tactics that massively increased their numbers so that the possibility for interaction increased. Finally, the parasitic way of life required the evolution of some mechanism for getting into the host and usurping the host's metabolism to create a new generation of parasitic offspring. Internal parasites, to be successful, also had to adapt by producing a variety of countermeasures to prevent the host cells' defenses from killing them.

Later parasites evolved sophisticated and complex life cycles to increase the probability of interaction between the parasite and the host. These included using one or more intermediate hosts in which the parasites could multiply enormously, as well as intermediate predatory hosts which were mobile and equipped with sensors that allowed them to find their prey (the final host), and mechanisms to introduce the parasite into the final host.

The First of Many

The earliest parasites were in all probability viruses that preyed on ancestral bacteria. While we have no fossil record to justify this assumption, it seems reasonable. Modern-day bacterial parasites, the so-called bacteriophages, can serve as a model. There are many known species of bacteriophage viruses. They look like miniature lunar landers, having a hexagonally shaped head which consists of a protein envelope filled with viral hereditary material, a collar, a tubelike penetrator, and an array of long, leglike structures. The tips of these legs bear recognition proteins that are highly specific for molecules of bacterial surfaces. Once the phage has landed and is locked in place on the bacterium, the penetrator is driven in and, like a hypodermic needle, it injects the viral genetic program into the bacterium. Inside of the bacterium, the viral genetic program takes over. Using the host's metabolic machinery, the phage begins to manufacture new viral materials, which in turn are assembled into new bacteriophages. In some cases, the host releases the new bacteria over a protracted period of time. In other cases, the host bacterium, filled with newly constructed phage, ruptures, releasing a horde of new, infective phage particles.

To Kill or Not to Kill

The ideal strategy for a parasitic way of existence differs from that of other predators in that parasites usually do not destroy their prey. It is clearly advantageous to keep the prey, the host organism, alive, to provide the parasite with lodging and board, while at the same time producing only minimal damage to the host. Using such a strategy, a parasite can produce vast numbers of offspring, thereby increasing the odds that its species will be preserved. For example, a single pair of schistosome worms can live in its

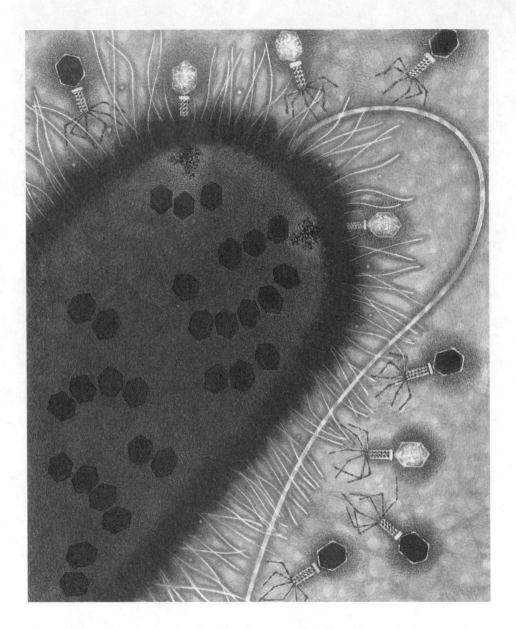

Getting a Life Bacteriophages are viruses that attack bacteria. Like all viruses, they are incapable of living on their own and require a living host's metabolic machinery to reproduce. Incapable of powered movement as well, they rely on chance encounters with the right kind of bacterium. When contact is made, special identification proteins on the bacteriophage's appendages latch onto the bacterium; then a hypodermic-like penetrator pierces the bacterium's armor and the virus's genetic program is injected. Once inside, this program takes over the host's own genetic program to make new viruses, eventually killing the bacterium and spilling out a hoard of new infective particles.

human host for dozens of years, producing a fertilized egg every few seconds. In the course of a decade it can produce hundreds of billions of eggs. However, not all parasites are successful in keeping their host alive and end up killing themselves in the process.

Adaptation to the Parasitic Way

We big-brained humans have tended to look at parasites as degenerate life-forms evolved from some free-living ancestor. The reason for this viewpoint is that in the process of evolving to the parasitic way, many structures that were not adaptive either shrank or were totally lost. Structures such as eyes, ears, big brains, motor control centers, even bowels often disappeared. After all, why keep a whole bunch of machinery that is useless to your survival? But, parasites are not simple, degenerate life-forms. They are highly evolved specialists whose adaptations have been so successful that they now outnumber free-living species or hosts by more than two to one.

Surface-Dwelling Parasites

Among the most annoying, innocuous, and harmful parasites are the ectoparasites, those specialists that spend some if not all of their life on the surface of their host. The host's surface provides a veritable cornucopia of food as well as a permanent home densely populated by potential mates. The appearance of warm-blooded creatures with feathers or hair provided a tremendous impetus for the evolution of new, more efficient ectoparasites. Furthermore, since birds and mammals are programmed for occasional intimate contact, the ectoparasites were provided with frequent access to new hosts. We will examine only a few of these ectoparasites, concentrating on those that have had profound and often disastrous impacts on humans. These are fleas, lice, mites, and ticks.

On Being Lousy

Of all blood-feeding organisms that view humans as a free lunch, fleas and lice have proven to be the most injurious, often altering the course of history, particularly in wartime. In biblical times, the Assyrian king, Sennacherib, led a great army southward to invade the land of the Hebrews. Victory by the invaders seemed certain until some mysterious plague decimated the ranks of the Assyrians and they were forced to return home defeated. This defeat was later recounted by Lord Byron in his poem, "The Destruction of Sennacherib":

> *The Assyrian came down like the wolf on the fold,*
>
> *And his cohorts were gleaming in purple and gold;*
>
> *And the sheen of their spears was like stars on the sea,*
>
> *When the blue wave rolls nightly on deep Galilee. . . .*
>
> *For the Angel of Death spread his wings on the blast,*
>
> *And breathed in the face of the foe as he passed . . .*
>
> *And the might of the Gentile, unsmote by the sword,*
>
> *Hath melted like snow in the glance of the Lord!*

In all probability the "Angel of Death" was the wingless body louse infected with a typhus-causing bacterium.

Later, Alfred Lord Tennyson wrote a poem about the Crimean War (1854) in which, again, more deaths among the combatants were due to disease than to enemy action. At that time, however, the cause of the disease was not known. Apparently, the disease had some preferential component: In all wars, French soldiers seemed to be more susceptible than soldiers of other nations. In France's greatest military debacle, Napoleon Bonaparte's attempted invasion of Russia, louseborne disease was a far more important factor than the Russian winter or the bravery and strategies of the Russians. Napoleon started out with almost two-thirds of a million men, but, when the army crossed into Poland, a massive epidemic broke out and continued to ravage the French all the way to Moscow and back. Finally, the depleted French were forced to retreat; less than 3,000 made it back to France. The victory was celebrated by Pyotr Ilich Tchaikovsky in his *1812 Overture,* which could have been more appropriately entitled "Hail to the Louse," because it was the human body louse, transmitting typhus, that did in the French!

Lice are blood-sucking insects that are very host-specific ectoparasites of warm-blooded vertebrates. They are wingless, armored, and flattened dorsoventrally and have legs equipped with terminal grasping claws that fit the dimensions of human hair. They spend their whole life cycle on or near the host in clothing or bed clothes. The sexes are separate, but males and females have no trouble finding each other. The female glues her eggs to the base of a hair, and development takes place in these so-called nits. Primates, particularly humans, attempt to remove them in a laborious, mindless, repetitive effort; this is why the phrase "nit picker" is given to those indulging in such behavior, for example, academic administrators, government bureaucrats, and IRS accountants.

Humans bear the dubious distinctions of serving as a feeding platform for no less than three different types of lice: the head louse, *P. capitus;* the body louse, *P. humanus,* which is closely related and can interbreed; and the crab louse, or pubic louse, *P. pubis*. The latter, as its name implies, is most often found on the course, textured hairs of the pubic areas, although it can also get into eyebrows, beards, and moustaches. Its transmission from one human to another usually occurs during sexual encounters when lovers are in close contact. Although such intimacies are relatively rare, they are frequent enough to ensure that new hosts are always available.

Epidemics of head lice periodically occur in today's highly sanitized world, even in relatively affluent communities. *P. capitus* is small and somewhat elongated. It is usually found in school children, as it holds onto the fine hairs of their scalp with its viselike claws, whose grip size matches that of the hair. But it is the body louse of *P. humanus* (or cootie as it was named by the soldiers of World War I) that is the real villain. Unlike other lice, the body louse can spend a lot of time off its host, residing in the host's clothing. For this reason, lice

Napoleon's Nemesis   The common body louse is a bloodsucker and also a carrier of typhus. The louse-ridden forces of Napoleon's army were decimated by this disease during their 1812 campaign through Russia.

have been referred to as "seam squirrels," and thousands of lice and nits can be found in the clothing of a heavily infested person. If these lice feed on a human infected with typhus, they pick up a bacterium, *Rickettsia prowazeki,* which infects the louse's bowel and kills it. But in the process of dying, the louse's feces act as the source of new infections. The *Rickettsia* are among the smallest of all bacteria and are cellular parasites. They are extremely potent pathogens and are frequently lethal. The lice and typhus flourish where people are crowded together under adverse sanitary conditions such as soldiers, prison inmates, and slum dwellers. The last war to give rise to the ravages of louse-borne typhus was World War I. Since that time, the discovery of potent insecticides to control lice and the development of anti-*Rickettsial* antibiotics have made typhus a minor player in the history of human conflict.

Rats, Fleas, and the Plague

The wonderful jumping capabilities of fleas were covered in detail in Chapter 5 on aerial predator-prey relationships. All fleas jump; here we focus on only one of the 1,500-odd species of fleas: the rat flea, in particular the oriental rat flea,

Microscopic Death   The tiny, bloodsucking rat flea, infected with the plague bacterium, transmitted this devastating disease to humans many times during the Middle Ages. In one great pandemic, one-third of the entire population of Europe was wiped out.

Xenapsylla cheopsis. It feeds on many different species of rodents but is most frequently encountered in black rats. Unfortunately, these fleas frequently abandon their rat hosts and feed with enthusiasm on humans. The fleas themselves are victims of a deadly bacterium, *Yeserina pestis,* which they get when feeding on an infected rat. Once in the flea's bowel, the bacteria proliferate rapidly, blocking the bowel so that the flea regurgitates masses of bacteria onto the rat, which in turn becomes infected and dies. The surviving infected fleas abandon the rat and seek a new host, which is frequently a human. Humans and rats began living together early in human history. Humans create warm habitats and are messy homemakers, inadvertently attracting rats to their leftovers.

Humans bitten by fleas afflicted with *Y. pestis* develop symptoms within several days. These are followed by the appearance of hard, swollen, purplish bumps, or buboes. Fifty percent or more of those stricken will die from the Black Death, or bubonic plague. Today, we have potent antibiotics that prevent plague epidemics in humans, although *Y. pestis* is still present in some rodent populations. In the past, vast pandemics (major outbreaks) spread over much of Asia, Europe, and North Africa. The Black Death started in Asia and the pandemic spread to Europe. In the mid-1300s the plague killed vast numbers of people and permanently altered the social order, the culture, and even the political and religious power structure of Europe. The first great European pandemic killed about one-third of Europe's population. A second pandemic in the mid-1600s, the great plague of London, wiped out 80 percent of London's population. The last pandemic, in the late 1800s, killed 8 to 10 million in the Far East and spread to Africa, Hawaii, and North and South America, where, fortunately, it had relatively little impact. Although there are still cases of plague, the possibility of global pandemics no longer exists.

Ticks

One group of opportunistic hunters that have become successful predators are the ticks. Ticks are distributed worldwide and feed on all vertebrates except fish. They have few natural enemies, can regenerate lost body parts, and can survive for as long as four years without feeding. Males, females, and all larval stages feed on blood and seem to have an affinity for mammals, including humans. Like spiders, scorpions, and mites, ticks are arachnids. They have four prominent pairs of legs, each of which terminates in two curved, pointed hooks. Their bodies are covered with a tough, leathery, external skeleton which provides them with a quite good armored protection. Their feeding apparatus consists of a pair of small retractable jaws, a pair of sensory appendages, and a central probe fitted with backward curving teeth; all of these are attached to a platelike structure, the hypostome.

Ticks are programmed to climb. Once on their perch, perhaps a blade of grass, they assume an attack position with the forelegs extended. They have an array of sensors to detect potential prey and are able to sense the presence of a mammal from as much as 25 feet away. Their eyes can detect shadows, and they can sense vibration, touch, and odor as well. They grab onto the prey with their forelegs and immediately begin to search for a good feeding area where they anchor onto their host, imbed their hypostome into the skin, make a painless tear, and begin feeding on the host's blood. To fortify attachment to their prey, they also secrete a glue that cements them in place. Feeding may last from hours to days, and the females' bodies balloon as they fill with blood. Eventually, when sated, they let go and fall off the host. Their reproductive capabilities are prodigious, and the large number of larvae that emerge can produce heavy infestations. One acre of grassland can house thousands upon thousands of ticks.

Unfortunately, ticks can become infected with a large variety of disease-causing organisms and, in the process of feeding on mammals, transmit these pathogenic organisms to their hosts. Among the diseases vectored by tick bites are a *Rickettsial* bacterium that causes Rocky Mountain spotted fever, two spirochete bacteria that cause Lyme disease and relapsing fever, a virus that causes La Crosse disease, an influenza-like microbe that causes Q fever, and even a protozoan parasite that causes babesiosis.

Host Countermeasures to Ectoparasites

Surface-dwelling ectoparasites are susceptible to attack not only by the host organism but also by cooperative symbiotic species, such as the small, cleaning wrasse fish that pick parasites off large predatory fish. In this typical case of symbiosis, the predator not only provides the wrasse with food but also shows appreciation by not eating the wrasse. In Africa, most vertebrate inhabitants of the grasslands are covered with ticks, fleas, and other ectoparasites, but they, too, have their little helpers, the birds called oxpeckers that swarm over their surface, even reaching into the animals' mouths, ears, and nostrils to gorge themselves on the pests. Another example of interspecies cooperation can be seen in the scaly anteater, the pangolin. This heavily armored creature is one of the few warm-blooded animals free of ectoparasites; this is probably due to its hunt for ants. When a pangolin attacks an ant nest, the defenders swarm over the invader in huge numbers. Unable to get through the pangolin's armor, the ants vent their wrath by killing any ectoparasite they find on the pangolin's surface.

Finally, ectoparasites of primates have to contend with yet another predator, namely another primate that is a member of the parasitized animal's social group. Watching a group of primates in the wild, one is struck by the enormous amount of time spent in grooming behaviors. Countless hours are taken up with meticulously inspecting the skin for parasites, picking the parasites off each other, and destroying them. Indeed, some students of primate evolution suggest that the development of an opposable thumb and fingernails, instead of claws, occurred to combat ectoparasites.

I've Got You under My Skin

Another group of skin parasites evolved a strategy that spared them the countermeasures used against exposed surface-dwelling ectoparasites; they became borrowers that tunneled under the skin. Practiced by parasites in both terrestrial and aquatic environments, this burrowing antipredator strategy has proven to be an effective survival tactic. Among the skin burrowers and tunnelers are the larvae of a number of flies whose species survival plan is to provide their larvae with a secure niche for development and an endless supply of nutrients.

My introduction to these parasitic larvae came when I was a visiting professor in Kenya. We lived out in the bush and had very few of the appliances we had become so dependent on back in the United States. There were no washing machines or dryers, and laundry was washed by hand and then laid out on the lawn to dry in the hot equatorial sun. Later, the dry clothes would be gathered up by the household staff and pressed with a very hot iron. They ironed everything, including socks and underwear.

At first, I thought this was a peculiar tradition, left over from the early African colonists. But, there was a method to this madness. The African tumbu fly, *Cordylobia anthropophagia,* lays its eggs on clothes; these eggs then hatch into invasive maggots that burrow into the skin, feeding on human tissues while completing their development. The extreme heat of the iron, however, is sufficient to kill the eggs.

Another fly, the bot fly, *Dermatobia humanis,* evolved a more-sophisticated strategy to get its parasitic larvae onto and into a human host. It employs a female mosquito or one of a variety of other blood-sucking anthropods as a hired gun. It plasters its eggs on these mobile hunting anthropods. There is no cost to the hunter, and when the blood-sucker alights on the skin of its warm-blooded prey, the heat causes the bot fly larvae to drop off and enter the hole in the skin barrier made by the piercing mouth parts of the mosquito. Over the next two to three months, the larvae feed and grow into inch-long maggots before emerging. Needless to say, such invasions are painful and irritating.

Another denizen found under the skin is a burrowing flea, the chigoe or sand flea, *Tunga penetrans,* which is common in South and Central America and Africa. The adults are found in dusty soil and, like their surface-dwelling cousins, are good jumpers. Their favorite targets are the feet, buttocks, and groin, all well within their jumping range, since people of these areas often squat. The female chigoe burrows under the skin with its egg-releasing rear end protruding to the surface. The eggs eventually fall off, develop, and produce new generations of chigoes. In all, over 100 species of insects have adapted part of their life cycle to parasitize under the skin.

Skin-Burrowing Mites

The other major group of organisms that exploit the vast feeding grounds under the skin of some hapless host are the mites. These creatures are cousins to the spiders, scorpions, and ticks. One such mite, that may effect as much as 75 percent of the human population, is the hair follicle mite, whose free lunch is provided by the oily sections of the sebaceous glands associated with the hair follicles. These parasites usually don't produce any symptoms and consequently are ignored. Not so with the mites belonging to the family Trombicularia, one of the 35,000 species of mites. The most annoying of these is *Eutrombicula alfredduges* (or chigger, jigger, or red bug) which has eight legs and a pair of piercing jaws. Chiggers have a complicated life cycle and live in dense, highly localized groups in grassy fields. Any intruder entering such a concentration will be heavily attacked. Only one larval stage hunts humans. These tiny larvae, only 1/150 inch long, move onto the skin and find a place to burrow, usually where clothing is tight. The chiggers chew into the skin and inject their salivary juices, which digest skin cells, turning them into a superrich frappé. The liquefied skin is then sucked up. As the chigger bores deeper, it forms a tube, called a stylosome.

After several days in residence, the chigger departs, leaving the tube filled with its irritating droppings. So, even after the chigger has gone, there remains an itchy reminder of its former presence.

However, the master arachnid denizen of the subcutaneous (under-the-skin) tissues is the highly contagious itch mite, *Sarcoptes scabei,* whose common names include the "seven-year itch" and the Norway itch. These mites are transmitted from host to host by casual contact and burrow under the skin, favoring sites in the groin and behind the knees, fingers, and toes. The female not only feeds and dwells in these meandering burrows but also lays its eggs and dumps its excreta under the skin. The host's body produces a severely itchy, irritating, allergic response to these foreign molecules. But, not all invaders of the dermis are anthropods; some are lower down the evolutionary scale. These are the roundworms and the flatworms.

The Flatworms

About 1 billion years ago, one of the earliest groups of multicelled organisms to appear was a relatively simple creature that lacked a body cavity and had no mechanism for transporting oxygen to its inner tissues. This being the case, evolution dictated a body plan in which all cells had to be near a surface that could supply oxygen, in other words, a flattened body form. These animals, the platyhelminthes, flatworms' early ancestors, were free-living marine and freshwater organisms that fed on debris and carrion. However, the flat basic body plan easily led to a parasitic way of life. Some evolved as external parasites, and some as internal parasites of vertebrates. Today, there are over 25,000 species of known flatworms. Many of them are parasites that afflict humans. The parasitic flatworms can be broken down into two groups: the trematodes, or flukes, and the cestodes, or tapeworms. No group of parasites demonstrates the essence of the parasitic way better than the trematodes.

The Flukes

The life cycle of the trematode parasites is more complex and evolved than almost anything else worked out in the process of natural selection. It involves two or more hosts, a variety of complex larval forms, and an extraordinary reproductive capacity that utilizes both sexual and asexual reproduction. One trematode group, known as the schistosomes, produces a debilitating infection that afflicts 1 out of every 30 people in the world, over 400 million cases, producing a widespread endemic disease called bilharzia (after Dr. Theodore Bilharz, whose studies first detailed the disease in Africa). The three major human schistosome parasites are *Schistosoma japonicum, S. hematobium,* and *S. mansoni,* although another large group of schistosomes of waterfowl do invade the human skin, causing local inflammation and itching but no parasitic infection. While the schistosome strategy calls for the infection to spare the host in order to allow the worms to spend years producing new potential offspring, they often weaken the host so severely that it becomes susceptible to the ravages of malnutrition and other diseases. The net result is that close to a million people a year die from schistosome-related illness. The details of the life cycle of the schistosome pose a which-came-first dilemma—the chicken or the egg. Let's begin with the egg.

The Intermediate Host   The life cycle of schistosomes requires that their larvae make chance contact with their adult host. Evolution has worked in such a way that the odds of encounter were increased by the use of an intermediate host—an aquatic snail—to produce what is known as a force multiplier effect. Hatched schistosome eggs develop into larvae that invade the snail and produce a huge progeny of another kind of larva, the fork-tailed cercaria (shown here swimming). Cercaria can detect human skin in the water, and then home in and penetrate it to start a new life cycle.

The adult female produces several hundred eggs per day, which are released into the host's bloodstream. But, to assure completion of the life cycle, the eggs have to get out of the host's body and into freshwater via either the urine or the stool. We know that the eggs can penetrate first the blood vessel walls and then the multilayered walls of the bowel or bladder. Most don't make it and end up being attacked by the host's defenses, which surround them and form cysts. The cyst is both a reproductive dead end for the parasite and the cause of a debilitating inflammatory reaction in the host. Eggs that leave the body and end up on soil, in a sanitary system, or in saltwater also represent a reproductive dead end. But those eggs which are released into freshwater start to absorb water rapidly, due to osmotic pressure. Then, they swell and burst, releasing a tiny, motile, ciliated larva called a miracidium.

Some miracidia are eaten; others swim in search of a host, namely a fresh-water snail. Not just any snail will do. The miricidia are genetically programmed to be highly selective for a given snail species, although all snails produce an array of chemicals that cause the miracidia to move toward them. If the miracidia enters the wrong species of snail, that too is a terminal event, but if within a half-day (the life span of a miricidium) the larva penetrates the correct snail, the next phase of the schistosome's life cycle begins. The miricidium changes form and becomes an internal parasite of the snail. While this infection in all probability shortens the snail's life span to about half a year, it allows the schistosome to generate hundreds of thousands of new larval forms called cercaria. As many as 1,500 cercaria a day are released back into the water.

There are two forms of cercaria, males and females, although how this divergence occurs is a mystery. The cercaria have a pair of suckers, some sort of penetrator mechanism, and a chemical homing system that is exceptional. Each cercaria is about 100 microns long (25,000 microns equal 1 inch), with about two-thirds of that length consisting of a muscular tail that is forked at its end (but note that not all types of cercaria have forked tails). Since the cercaria have no feeding apparatus, they have only a limited time, about one day, to find an appropriate host, that is, one that gives off an attractant. Despite the lack of noses, eyes, and other elaborate sensors, it has been shown that cercaria use chemically cued movement, sight, and touch to find their prey. The initial signal that stimulates their search and attack is the presence of the amino acid arginine in the water. However, for the attached cercaria to penetrate, they must confirm that they have found a specific target. This information is provided by a mixture of fatty waxes and oils secreted into the hair shaft by the sebaceous glands of the host.

Once the host's identity has been established, the cercaria plunges into the hair shaft, sheds its tail, and then penetrates into the host by some unknown process, probably enzymatic digestion. Meanwhile, the tail of the cercaria, which is rich in arginine, acts as an attractant to other cercaria, a sort of invitation to come and mate, since a single cercaria can't reproduce without benefit of a mate. After cercaria penetrate the final host, they get into the blood vessels. How they do this is another mystery. Once in the blood vessels, they undergo yet another change in form, becoming male or female adult worms.

Adult worms navigate in the stygian darkness of the blood vessels, finding their way first to the lungs, an oxygen-rich environment, and then to the liver. The liver is the most nutrient-rich environment in the body, because all food-stuffs absorbed by the bowel are transported to the liver via a specialized anatomical arrangement called the hepatic portal system. What signals the worms to position themselves in these particular blood vessels is yet to be resolved, but when they reach their destination, the heavy bodied male and

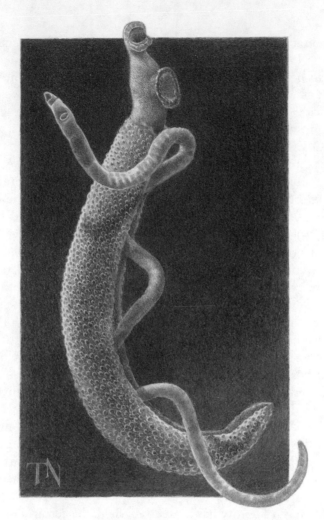

In Flagrante Delicto for Life   Finding a mate in the darkness of blood vessels is a daunting task. Male and female schistosomes locate each other by way of chemical cues, and then the male enfolds the female. Thus begins a never-ending orgy that can, within a schistosome's lifetime, produce millions of eggs.

slender female attach to the vessel walls with a pair of suckers. The males then release lipid pheromone, a perfume that attracts nearby females to come hither and embrace. The male has a pair of fleshy folds on its ventral side, and the female is permanently embraced in these folds, with female parts and male parts in a state of flagrante delicto for the lifetime of the worm, which can be up to ten years.

After a week or so, the female starts to produce eggs, and recent research has shown that the schistosome has by now usurped the host's defense system to stimulate its own egg production. The immune system, part of the body's inner defenses, involves an elaborate collection of chemical signals which activate cells to attack the invader. The schistosomes apparently use one of these chemical response signals to stimulate egg production, once again increasing the odds that their species will survive.

What an incredible wonder it is that a whole series of highly improbable events in the schistosome's life succeed in preserving the species. Life is like a game of craps, and usually highly improbable events simply don't occur, unless there is a vast number of chances or trials. The individual's odds of hitting the lottery by picking six or seven digits is about 1 in 50 to 100 million, but there are always winners. The odds of a schistosome pair's producing another breeding pair may be 1 in 1 billion. A bad bet? No, not when one considers that in a ten-year lifetime, a pair of worms produces a half trillion eggs. Given that number of possibilities, even highly improbable events become a certainty.

The Tapeworms

The other class of parasitic flatworms is the Cestoda, a long, flat, noodle-like parasite, whose size is limited to the length of the bowel of the host. One tapeworm found in whales is over 100 feet long. Structurally, the adult worms are very simple organisms. They are devoid of eyes, mouths, brains, and all that sophisticated biological hardware needed by free-living organisms. What they do have is a head equipped with a ring of hooks and four suckers used for attachment, and a regenerative zone that continuously produces reproductive segments throughout the lifetime of the worm. Early in its life, each reproductive segment develops a complete set of both male and female parts. These segments, called proglottids, can't fertilize themselves, but they can interact with other proglottids flopping around in the bowel to cross-fertilize each other. How this is done is unknown, but it happens with great frequency.

The proglottid continues to grow, discarding unwanted parts until it contains thousands of eggs. With each bowel movement, hundreds of egg-laden proglottids are shed into the water. Now the game of finding a new host begins. To do this, there are two strategies used, one terrestrial, one aquatic. The complex aquatic life cycle involves several hosts in producing one of the largest human tapeworms, *Diphilobothrium latum*. Classifiers keep changing the names, so we will use its common name, the broad fish tapeworm.

The adult worm, generally resistant to the host's defense system, dwells in the small intestine, immune to the bowel's digestive enzymes. The surface of each proglottid has thousands of tiny projections, which absorb digested nutrients. Unless infected with a huge number of worms, the host, a human or often a bear, suffers no ill effects. My parasitology professor had one such tapeworm (which he called Max) and, thus, always had a supply of fresh proglottids for class use. If the proglottids reach freshwater, they burst, releasing huge numbers of fertilized eggs,

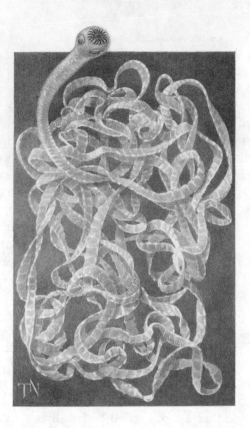

100 Feet of Parasite   Tapeworm larvae, when ingested, develop into long, noodle-like, flat worms that absorb their nutrients from the host's bowel. These worms attach to the lining of the host's bowel with a ring of hooks and four suckers, and then begin to generate segments that are reproductive units. A tapeworm can produce hundreds of millions of eggs in the course of its lifetime. The longest tapeworm, found in the blue whale, is 100 feet long.

which develop into free-swimming larvae. The larvae are then eaten by a copepod, a tiny aquatic relative of shrimps and crabs. In the copepod, the first larval stage develops, but unless the copepod is eaten by a fish, the cycle ends there. In the fish, the third larval stage develops within the muscles, and this third stage can be transmitted from small fish to the large fish that feed on them.

Humans can then get the worm by eating uncooked freshwater fish. Fortunately, most sushi dishes use only saltwater fish, but there is one segment of the human population that gets infected by the broad fish tapeworm. Jewish women, when making a traditional dish called gefilte fish, chop the raw fish, add condiments and then taste before cooking to make sure the flavor is correct and, in the process, may acquire a tapeworm infection. However, observant Jewish and Moslem people never get the pork tapeworm, since pork is forbidden by their religious laws.

The pork tapeworm, *Tenia Solium,* can only mature in the human digestive tract. This represents an amazing example of parasite-host specificity. The eggs are passed in the stool and eaten by pigs. In the pig's muscle, the worm develops into a cyst. Eating undercooked pork containing such cysts allows the cycle to continue.

The Parasitic Roundworms

Millions of years after the flatworms came into being, the roundworms (nematodes) evolved. These animals had a complete bowel from mouth to anus, plus a false body cavity, and were destined to be one of the most successful of all animal groups. Today, only two groups on earth have a biomass (total weight) greater than that of humans; these are the insects and the roundworms. The ancestral roundworms were free living with a variety of feeding strategies, but a number of them adapted to a parasitic way of existing. Some of these afflict humans, causing a variety of debilatory or even lethal diseases.

Like all parasites, the roundworms multiply rapidly and prolifically, producing tens or even hundreds of thousands of eggs a day. Some are accidental residents of humans who eat sushi or ceviche, and thereby interrupt the normal cycle of the parasite. That cycle starts with the eggs' being eaten by small, shrimplike krill, which are, in turn, eaten by fish and then develop in fish muscle. The final host of this roundworm, *Aniakis,* is usually some fish-eating marine mammal, but when humans disrupt the cycle, the parasite bores into the human's bowel wall, causing agonizingly painful, gut-wrenching spasms. However, there are also a number of parasitic roundworms that directly afflict humans by entering their mouths, boring through their skin, or being transmitted by a blood-sucking insect.

Over 20,000 species of roundworms have been described. Most are free-living, soil-dwelling forms, but a multitude of nematodes have adapted to a parasitic mode of life. They range in size from the microscopic filarial worms that live in small blood vessels to a giant, 30-foot-long roundworm parasite found in the placenta of the female sperm whale. We consider only a few types of parasitic roundworms that are parasites of humans: members of the genus *Strongyloides* and the hookworms, genera *Ancylostoma* and *Necator.*

Hookworms

Hookworms are among the most common and important human parasites, afflicting about a billion people, particularly in developing countries. While the symptoms of hookworm infection are not particularly spectacular, their impact can produce a workforce of low vitality and productivity. Many stereotypes of lazy, slow-moving lethargic southerners derive from the energy-sapping hookworms that lived in their bowels. Of the two species of human hookworms, we focus on the *Necator americanus,* which was introduced into the Americas during the slave trade.

The adult hookworms are about a half-inch long, with the females' being a bit larger than the males. Adults are found latched onto the walls of the small intestine of their hosts (humans and sometimes dogs). Their holdfast organ, a remarkable mouth cavity equipped with wicked, hooklike ventral and dorsal teeth, allows them to dig in and feed on blood. A heavy infection, 20 or more worms, removes iron from the host, causing iron deficiency anemia. It also can cause a protein deficiency, disrupted growth, mental retardation, and even death. There are effective medical treatments for the parasite, but the best control is a sound sanitation system which prevents the eggs from being deposited on sandy soil.

The female hookworm produces a prodigious number of eggs, which pass out with the feces and quickly develop into larval worms in warm, moist soil. These larvae are free living, feeding on bacteria and doubling their size, but as development continues, they stop feeding and become infective. They can survive for up to six weeks in the top few inches of soil. They are equipped with heat sensors, and when a warm foot alights on them, they enter the skin through the pores, hair follicles, or a cut in the skin, producing a localized inflammation called ground itch. The larvae then migrate into blood vessels, get into the air sacs in the lungs, and are coughed up and swallowed. In this way, they reach their final home base in the small intestine.

Unfortunately, man's best friend also is a host to hookworms, and dogs always deposit their stools on the ground, where infective larvae develop and can get into barefooted humans. Dogs also lick their rear ends, not to mention the rear ends of other dogs, and then affectionately lick the faces of their masters; in so doing, they can infect their masters. Sometimes the larvae fail to find a vessel and instead migrate through the skin, causing an inflammatory, itchy condition called a creep.

The Threadworms

A number of thin roundworms of the genus *Stronglyoides* may, in part of their life cycle, become human intestinal parasites that pose a significant threat to public health. One *S. stercoralis* in its parasitic phase infects tens of millions of people worldwide: 21 million in Asia, 4 million in Africa, 1 million in the former Soviet Union, and another 5 million in the Americas. They thrive best in those warm, moist climates in which sanitation practices are poor and people often go barefooted. Dogs and cats can also be infected with threadworms. In this way, pets act as reservoirs for human infection.

The adult female, which burrows into the lining of the small intestine, can reproduce without benefit of a male worm and produces large numbers of eggs which eventually develop in soil contaminated with the infected feces. In the soil, they undergo several stages of development and eventually molt into infective larvae. However, sometimes the intestinal larvae bore through the gut wall and enter the bloodstream; this enables them to continue their parasitic lives without leaving the host. Such autoinfections can give this parasite the capacity to maintain its cycle for 40 or more years and, in so doing, can produce heavy, increasingly debilitating infections. The more common pathway involves larvae in the soil penetrating the skin of the foot, entering the bloodstream, and proceeding to the heart and lungs, where they molt and create a local inflammation. They are then coughed up and swallowed, in this way getting into the small intestine. Heavy infections cause pain and intense burning sensations in the gut, accompanied by nausea, vomiting, and diarrhea, which lead to weight loss. The massive invasions of the tissue by microscopic larvae can be lethal, because the parasites' countermeasures to the host's defensive systems include immunosuppression similar to that produced by the HIV virus, the causative agent of AIDS.

The Hired Guns

A number of parasites devised clever tactics to perpetuate their life cycles. One strategy employed a secondary host that was a predator equipped with an array of hunting tools and endowed with mobility. These hired guns were often insects, and a multitude of human parasites are transmitted by insect vectors. Some were viruses, some were bacteria, some were parasitic roundworms, and several were single-sized protozoans. We consider three of the hired guns here: mosquitoes; their cousins, the biting flies; and the assassin bugs, since they are so economically important.

The Tsetse Fly and the Assassin Bug

Trypanosomes are flagellated protozoan parasites of the blood; they require blood-feeding insects as vectors to complete their life cycle. Once in the human body, the trypanosomes can assume a number of disguises, but the most viable form is a microscopic, long, slender cell with a single whiplike flagellum. There are two major groups of the genus *Trypanosoma* that cause human disease, one exclusively found in the Americas and the other exclusively found in Africa, where it causes the insidious disease called African sleeping sickness.

The two kinds of African trypanosomiasis are transmitted by no less than 21 species of biting flies of the genus *Glossina,* the most familiar of which is the nefarious tsetse fly. The flies pick up the parasite when feeding either on an infected human or on some wild animal that is immune to the parasite and acts as a reservoir host. In the fly, the parasite finds its way from the gut to the salivary glands and is transmitted to a new host by the female during feeding. Tsetse flies are extremely persistent hunters whose feeding apparatus, unlike the delicate stylets of the mosquitoes, is more like a bayonet or rapier forcefully driven into the skin, so that a profusely bleeding wound is produced. There is no subtlety to their attack, and from personal experience, I can attest to the sharp pain that accompanies the bite. Fortunately, only 1 out of every 100,000 flies carries the parasite.

Once the trypanosome infects a human, there is an estimated 50 percent rate of mortality, and those who survive are frequently left with permanent brain damage. Today, there are effective chemotherapeutic drugs that have helped keep the disease in check, but recent political instability in Central and East Africa has so disrupted medical services that sleeping sickness has reached epidemic proportions in some areas.

The American trypanosome was first described by Dr. Carlos Chagas in Brazil, and its manifestation bears his name, Chagas' disease. The protozoan is *Trypanosoma cruzi,* and its vectors are cone-nosed true bugs, also known as assassin bugs or kissing bugs.

The kissing bug, genus *Triatoma,* carries the parasitic trypanosome in its hind gut, and human infection occurs when the bug's feces get into the wound made by its mouth parts. These cone-nosed bugs can be quite handsome, with checkerboard orange and black markings on their wings. They range in size from ½ inch to a little over 1 inch and are nocturnal hunters. Their proboscis contains four needle-sharp, delicate lancets that produce a painless puncture. Vessel feeders, they prefer hunting on areas richly endowed with blood vessels, such as the lips. This is how the *Triatoma* came to be called the kissing bug.

Once in the human host, the trypanosome invades the cells of many tissues, but mainly the spleen, liver, and all types of muscle cells. Accumulations of the parasite in the heart muscle can be lethal many years after the initial infection. The disease is prevalent in South and Central America and affects more than 10 million people, particularly those living in primitive housing that provides daytime hiding places for the bugs. The bugs also attack raccoons, bats, armadillos, rodents, and household pets, who act as reservoirs for the parasite.

Mosquitoes

Mosquitoes are delicate, fragile, winged hunters, adept at transmitting a great variety of pathogens from submicroscopic viruses and bacteria to protozoan parasites and even some species of parasitic roundworms. These mosquitoborne diseases have had enormous human impact, killing hundreds of millions of people over the course of history. Even today, with all our medical advances, mosquitoes still cause millions of deaths and sicken hundreds of millions each year. There are 2,500 species of mosquitoes, ranging from the Arctic to the fringes of Antarctica. Some are found up to 14,000 feet in the Himalayas, and others thousands of feet below the surface in the shafts of mines. The disease vector is always a female, which after emergence from its larval state, is almost always mated. However, in the mosquito, the proverbial mandatory cigarette after sex is changed to a craving for a high-protein meal, usually in the form of warm blood. We considered in detail the hunting and feeding tactics of these blood-thirsty insects in Chapter 5; this section is devoted only to the life cycles of parasites that are transmitted by mosquitoes during feeding.

There are two particularly dangerous groups of mosquitoes, those of the genus *Anopheles* and those of the genus *Aedes*. The long-legged female *Anopheles* sports spotted wings and, when feeding, sits tail end up, at a 45-degree angle to its victim's surface. During feeding, she injects her saliva which contains anticoagulants, vasodilators, possibly a local anesthetic, and unfortunately, sometimes, the malarial parasites of the genus *Plasmodium*.

The mosquito pests most irritating to humans belong to the genus *Aedes*, which lays its eggs in any available water: swamps, vernal ponds, and any number of containers that trap rainwater. They are more robust than the *Anopheles*. Some are striped, and when feeding, all females hold their bodies parallel to the victim's surface. During feeding, infected *Aedes* can transmit a number of deadly viral diseases, including yellow fever, dengue fever, and several different forms of encephalitis. However, reports that the human immunodeficiency virus, HIV, the cause of AIDS, was transmitted by *Aedes* mosquitoes proved to be wrong.

The Viruses Vectored by Aedes

One of the most formidable viruses, yellow jack, or yellow fever, is vectored to humans by *Aedes aegypti*. Its name suggests an African origin. In any case, the slave trade introduced both the mosquito and the virus to the New World. It is a fascinating virus: within a few days, symptoms appear and then often seem to go into remission. However, during this time, the virus is devastating the liver, and soon the victim has internal bleeding and jaundice, a yellowing of the eye whites and the skin due to abnormal amounts of pigmented bile salts accumulating in the blood plasma. From 10 to 80 percent of those afflicted with this disease may die.

The cause of yellow jack remained a mystery until the Spanish-American War, when the disease killed eight times more U.S. soldiers than enemy action did. In a brilliant, bold piece of biological detective work, a team of U.S. army physicians soon established that the vector was a mosquito, and eradication of mosquito-breeding places brought epidemics under control. It was also found that victims of yellow fever who survived the ravages of the virus were now immune. This led to efforts to produce an antiviral vaccine, and finally, by the mid-1930s, Nobel Prize winner, Max Thiler succeeded, and yellow fever was, in large part, controlled. However, in some poor developing countries it still exists, and in those countries, thousands die every year from the virus.

Dengue Fever

Recently, the Asian tiger mosquito, *Aedes albopictus,* was inadvertently imported into the United States. These flying tigers breed readily in any water-filled container, are ravenous feeders, and unfortunately, are also vectors for several viral diseases, including dengue fever. Dengue fever is found worldwide in warmer climates. Its manifestations range from a mild array of transient symptoms such as respiratory and gastrointestinal distress to fever, severe headaches, and muscle pain, from which is derived the common name breakbone fever. More severe cases produce bleeding and shock. Fortunately, these are rare, but they do require emergency treatment, which is mainly in the form of supportive therapies.

Bad-Air Disease

Human history is rife with stories about a great variety of epidemics that killed huge numbers of people and altered the course of human affairs. One particularly insidious plague that was characterized by alternating high fevers and chills was referred to as a "burning Ague" in the Old Testament and as the "King of Diseases" in Indian writings. Hundreds of years before Christ, there was an indication that this disease was associated with the stagnant waters of swamps. People of that era dedicated a deity to this disease, whom they called Dea Febris, from which the word *febrile* (meaning "feverish") draws its origin. Later, the Italians presumed that the disease was associated with the bad air, or *mal aria,* emitted from the swamps. From this came the modern-day English name malaria.

By the late 1800s, a number of studies demonstrated that the bad seed was a microscopic parasite in the blood, and by 1884 it had been shown that blood from an infected person injected into healthy people could cause malaria. Camillio Golgi, who shared the first Nobel Prize in Biology and Medicine (in 1901) showed that the waxing and waning of temperature seen in malaria reflected its infected cycle and also demonstrated that different species of malarial parasites produced different febrile patterns.

The final breakthrough in understanding the malarial cycle was provided by an English physician, Ronald Ross. He was aware of the studies of Patrick Manson (who has the distinction of having had a schistosome worm, *Schistosoma mansoni* named after him), which suggested that the parasite was vectored by mosquitoes. Ross's work was able to demonstrate the various stages of the life cycle in both the female anopheles and in humans. He was awarded the second Nobel Prize in Biology and Medicine in 1902.

The malarial parasite ravages red blood cells, sucking them dry of hemoglobin in the process of reproducing asexually. Recent research has shown that the malarial parasites had a very ancient ancestor, going back more than 1.5 billion years. This ancestor was capable of photosynthesis and contained chloroplasts. The modern-day malarial parasites retain these relics of their evolutionary past, even though they are never exposed to sunlight. As we look at the evolution of free-living organisms to the parasitic way of life, we are struck by the disappearance of unused body parts over the course of time. But, the ways of evolution are such that some unused structures, although functionless, are retained.

All four forms of human malaria follow identical life cycles. The reproductive flagellated sex cells, which arose from the merozoites, swimming in the human blood stream are sucked up by the mosquito during feeding. In the mosquito's gut, the sex cells fuse, forming fertilized eggs which develop into long, thin sporozoites which find their way to the mosquito's salivary glands. This sporozoite-laden saliva is injected into humans during feeding, and the sporozoites reach the liver.

Once in the liver, the sporozoites metamorphose (change form) into merozoites and then develop into yet another form called schizont. The merozoites released from the liver cells now invade red blood cells and, feeding ravenously, undergo prolific reproduction. The red blood cells burst open, spewing out large numbers of merozoites, which start a new cycle by invading other red blood cells. The release of new merozoites follows a relatively precise time course and triggers a surge of fever followed by chills. Each of the four forms of malaria has its own schedule, which can be used in diagnosis. *Plasmodium vivax* and *P. ovale* have a 48-hour cycle, *P. malariae* a 72-hour cycle, and the deadliest of all the malarias, *P. falciparum,* a 36- to 48-hour cycle.

P. falciparum is responsible for about 80 percent of malaria worldwide and is resistant to the best and cheapest antimalarial medication, chloroquine. The parasite also produces the highest mortality rate of the four human malarial parasites. The destroyed erythrocytes, on reaching the spleen and liver, produce a dark-colored, iron-containing pigment and can clog and obstruct small blood vessels, including vessels in the brain. These obstructions can be lethal. The massive destruction of red cells can also dump enormous amounts of hemoglobin into the urine, making it dark. This characteristic has given *P. falciparum* the name blackwater fever.

There are medications that can treat *P. falciparum,* but the parasite is adaptable and mutagenic changes produce drug resistance. The search for new and better antimalarial therapies is ongoing, and given the tremendous advances made in modern biotechnology, we may yet defeat the parasite. But I wouldn't bet on it. As of this writing, malaria still kills almost 3 million people, mostly children, every year.

Chapter 7

How Plants and Fungi Make War

Predators of Plants

Plants have no means of escaping from their enemies, but they have over the course of 350 million years devised a myriad of vigorous defenses to counter the assaults of both vertebrate and invertebrate herbivores. There are over 350,000 species of insects that obtain their nutrients by eating plants. Add to this legion their eight-legged cousins, the spider mites, and you find that there are more species of vegetarian invertebrates than there are species of plants. Some chew on plants; some suck on plant juices; some burrow into the interior of plants; others are specialists in feeding on flowers, seeds, and fruits; and yet others subvert a plant growth hormone to produce galls, abnormal growths which house and feed their young.

There are also multitudes of vertebrates that prey on plants. Some of these are obligate herbivores; others are omnivorous, feeding on both plant and animal tissues. They are munchers, browsers, and fruit and seed eaters. One formula for evolutionary success as a herbivore was giantism, and given the poor nutritional value of a vegetarian diet, these huge animals had to consume large amounts of plant material to maintain themselves. For example, a modern elephant that weighs 5 to 6 tons eats about 500 pounds of food per day. The elephants' larger and only recently extinct predecessors, the mammoths, needed even more. During the heyday of mammalian evolution, 2 to 60 million years ago, there were 22-foot-long, elephant-sized ground sloths and rhinoceros-like monsters called titanotheres. These massive, small-brained ungulates stood 7 feet high at the shoulder, much bigger than elephants, and were leaf browsers. All of the titans became extinct about 32 million years ago when the earth's climate changed.

Prior to the mammals, during the Age of Reptiles, the size of giant terrestrial herbivores reached its peak. The largest of these animals, the sauropods, reached weights 20 times greater than those of modern elephants, and one called supersaurus was 150 feet long from the tip of its snout to the tip of its tail. Unlike mammals that burn calories to maintain a stable body temperature, these giants were cold blooded, so we can't really know how much plant material they ate; but clearly, the thunder lizards (brontosaurids), traveling in herds, could devastate the local greenery. Like all herbivores, their digestion of plant material was aided by bowel-dwelling microorganisms, some of which produced methane gas. As a matter of fact, modern ungulates and termites pass huge amounts of methane gas, which, along with carbon dioxide, is responsible for the greenhouse effect. Anyone who has been around cows or elephants for any length of time can attest to their flatulence. One can only imagine the eructations emitted by a 100-ton sauropod. Perhaps the name thunder lizard is appropriate in more than one way.

The devastation wrought by large, modern herbivorous elephants was apparent in Kenya's Tsavo National Park when the elephant population reached 40,000. Everywhere you looked, there were shattered trees; even the gigantic baobob trees were gouged almost to extinction. Today, there are only about 4,000 elephants in Tsavo and the balance has been restored. Although large herbivores have enormous ecological impact, small insects have a more staggering impact when their populations explode.

Insect Plagues

The insect's potential for reproduction under ideal conditions is staggering. Fruit flies can produce 25 generations in a year, and each female offspring lays a hundred eggs. Of course, not all survive, but theoretically, they can produce a trillion trillion trillion flies in a year. The most familiar example of massive insect population growth is that of the migratory locusts. Certain conditions periodically create enormous numbers of long-winged, flying migrants, which form huge swarms for weeks at a time and cover thousands of miles in their quest for food. They grow to maturity in the desert, and when conditions are right, shortly after sunrise they start streaming into the air, flying in the same downwind direction, and recruiting on their way millions of locusts. The streamway can cover 400 square miles, and each individual locust can consume its own weight in plant matter in a single day. Each square mile of them (and they number in the tens of millions) can eat 200 to 250 metric tons per day. Ever since humans began to cultivate plants for their own use, the darkening of the skies by a locust swarm has signaled disaster.

Both the Old and New Testaments made frequent references to sudden visitations by these orthopteran hordes. For their own selfish purposes, humans became allies of the plants to reduce the swarms. Today, early warning systems and the use of preemptive strikes with aerial spraying have limited the impact of locust plagues.

Caterpillars, such as the armyworms and gypsy moths, also exhibit periodic population explosions with devastating results. A recent gypsy moth attack in New England laid bare many forested areas. If you walked through the woods, you could hear their munching, and you were inundated with a constant rain of falling pieces of leaves and gypsy moth excrement. I was also witness to the invasion of armyworms covering the ground in my Kenyan farm. This abundance of food in turn attracted squadrons of birds to prey on the pests. But the plants under attack were also fighting back.

Chemical Warfare

Many plants manufacture and deploy a number of chemical weapons to defend themselves from attack. It is estimated that plants and mushrooms have an arsenal of over 100,000 compounds. Some are stored, so that the plant's toxic weaponry can be immediately deployed. Other defensive chemicals are manufactured on demand in response to damage. The synthesis response is slow and requires that the plant's genetic program be activated by some product engendered by the wound inflicted by the insect.

Communications and Signals

Sometimes the ability to communicate information is essential for survival. However, plants and fungi lack nervous systems, so they communicate by purely chemical means. The informational molecules are released and can be carried by the plant's vascular system, a system of tubes carrying sap, to send messages to other cells within the plant. We know that damage to even a few leaves can somehow activate that plant's genetic machinery to suddenly start manufacturing ever-increasing amounts of defensive chemicals. In a sense, the message, "Am under attack" warns other parts of the plant, "You may be next," and forewarned is forearmed. The exact chemical nature of these alarm substances is not fully known. But one experiment showed that cotton plant leaves under attack by a fungus released some volatile airborne signal that caused unattacked cotton plant leaves to expand their production of antifungal defensive compounds. If the warning molecule is carried by air, then is it possible that plants can warn nearby members of their own species of an impending assault?

The signal, "One if by land, two if by sea," displayed by the lanterns in the Old North Church tower, allowed Paul Revere and his fellow riders to broadcast the message, "The British are coming," allowing the minutemen to repel the British columns. Many plants have similar messages that signal their neighbors. For example, some species of willow send a warning, "The tent caterpillars are coming." The airborne message may not be completely species specific, and other species within the community may also read the signal and start their own weapons production to deter predators.

Airborne signals from plants can be read not only by other plants but also by insects. For example, undamaged lima bean plants give off an attractant that invites attack by herbivorous spider mites. However, following an attack on a plant, the juice-sucking mites alter the plant's signal in such a way that it now acts as a spider mite repellent, and thereby discourages other herbivorous mites from exploiting the lima bean plants. Not only does the plant under assault begin to release the volatile repellent, but nearby bean plants pick up the signal and also synthesize and release the repellent. The same airborne message reaches yet another species of spider mite, this one a tiny, ravenous carnivore that is carried by the wind. If, in their random flight, these predators land on a plant besieged by herbivorous mites, the chemicals released by the damaged plant say, "Be my guest; there is lots of food available," and the carnivorous mites gorge themselves on the herbivorous mites until the plants, no longer under assault, stop producing the chemicals that recruit the carnivores.

The apparent ability of a plant to communicate with predators that eat the herbivores attacking it is not all that unusual. Corn plants under attack by armyworm caterpillars send out chemical deterrents which also function as signals to a small parasitic wasp that attacks armyworms by depositing its eggs in them with its piercing ovipositors. The eggs develop into larvae, which then consume the armyworms from the inside. This raises the question: did plants evolve a specific signal system to recruit armyworm-killing wasps? Probably not; the wasp may simply be programmed to home in on any odor associated with the presence of armyworms.

Surface Defenses of Plants

Over the span of 350 million years of being munched on by vast armies of herbivores, plants also evolved an array of structural surface defenses that rapidly deter, repel, or kill would-be herbivores. Included in their armory are a wide variety of strong needle-sharp thorns, a number of needle-like, thin, toxic urticating hairs, and several types of trichomes that secrete adhesives.

Many of the larger herbivores are deterred from munching by plants equipped with thorns. Thorns of all sizes are remarkably sharp and exceptionally strong, rapier-like weapons that produce painful puncture wounds and

Getting Bogged Down   Surface defenses of plants include different types of microscopic hairs called trichomes. Trichomes secrete sticky, glue-like substances that are capable of immobilizing small, herbivorous insects such as the ant shown here.

scratches. Pain is an immediate aversive stimulus that causes the wounded animal to rapidly withdraw from the source, and once having experienced the plant's barbs, many herbivores learn to recognize and avoid them.

The stinging nettle, *Urtica dioica,* a weed, is covered with long, thin, hollow stinging hairs. At the base of each hair is an enlarged bladder containing a nasty mix of potent irritants, which have profound effects on blood vessel diameter and permeability. The leaves and stems are covered with the stinging hairs. On contact, these needle-like structures penetrate the skin, the bladders are compressed, and the irritant is squeezed out. The effect is almost immediate, creating an intense burning sensation, inflammation, and itchiness.

Very small herbivorous insects are dealt with by miniature surface weaponry called trichomes that cover the leaves' surfaces. Some are upright, short microscopic hairs topped by a rounded cluster of four glandlike components which, when mechanically stimulated, burst and release a sticky fluid. Other trichomes are taller and more slender; and these hairs continuously secrete another kind of sticky liquid which accumulates as small drops at the tip. Both tall and short trichomes are densely distributed over the surface of trichome-bearing leaves. With the sticky goo covering their legs, aphids and other small plant predators making contact with the tall trichomes soon find movement difficult, and also find the adhesive repugnant. The insect's struggles soon bring it into contact with the short, bulbous-tipped trichomes which then release their mix of chemical weapons. Once those have been released, the binary (two-part) chemical weapon is completed and, in the presence of air, forms a hardened mass that immobilizes the insect invader and eventually kills it. While not potent enough to immobilize larger and stronger insects with its adhesives, the trichome exudate still protects the plant from predation with other types of defensive chemicals that act to inhibit feeding.

Plant Poisons that Alter the Nervous System

Both vertebrate and invertebrate animals coordinate and control all their diverse behaviors by processing information in their nervous systems. Information is processed in circuits composed of specialized sensors which relay the information to integration neurons that analyze the signal and then relay that information to motor neurons that activate appropriate muscles to produce the desired response. The signals are bioelectric, rapidly conducted pulses that, on reaching the nerve ending, cause the release of messenger molecules called transmitters. In other words, when sensors talk to neurons and when neurons talk to other neurons and when neurons talk to muscles or glands, they do so by means of transmitter molecules. The messages are brief, a few thousandths of a second, and then the transmitters are either broken down or recycled. Over the course of millions of years of the coevolution of plants and plant eaters, plants evolved a variety of chemicals that somehow alter the functioning of the predators and, in so doing, devised some of the most potent nerve poisons known.

There are several ways to disrupt the process of neurotransmission. One tactic is to interfere with the manufacture of transmitter molecules. Another calls for somehow altering the release, breakdown, and recycling of the neurotransmitter molecules. Finally the plant toxins can either mimic or block the effect of the neurotransmitter on the next cell in the circuit. A variety of plants have developed poisons that can do one or more of the above to affect the herbivore's nervous system and subsequent behavior. The effects of such nerve poisons on humans have been known for millennia. Roman women, for example, used extracts of *Atropa belladonna* (Latin for "beautiful woman") to dilate the pupils of their eyes, thereby making themselves more alluring. A number of different tribes have used atropine-like plant extracts as arrow poisons, and atropine and its analogues have found wide application in modern medicine, in which it is used to dilate pupils for eye examinations, to prevent symptoms of the common cold, and as an antidote for nerve gas poisoning. All these effects involve blocking the normal activity of the neurotransmitter acetycholine.

The effects of plant-produced blocking agents (and of any other poison) depend on the dose, or how much gets into the body. Doses are measured in micrograms (millionths of a gram) or milligrams (thousandths of a gram) per unit of body weight (kilograms). In small doses, the effects may be medically useful, but in higher doses the effects can be either debilitating or lethal. For example, large doses of atropine can cause paralysis and death because atropine, like cobra venom, blocks the transmission of nerves to skeletal muscles. The nervous system has a number of different neurotransmitters that are affected by plant and mushroom poisons. It should be stressed that a poison's effects depend on how much is ingested and absorbed. With that in mind, along with the realization that many insects that attack plants weigh several millionths of the weight of the adult human, it is evident even the tiniest amount of poison can be effective in warding off plant herbivores. We consider here some of the plant and mushroom poisons for their effect on the human nervous system.

Since ancient times, it has been known that extracts of the opium poppy, *Papaver somniferum,* produce euphoria, block pain, induce sleep and dreams, inhibit coughing, and stop diarrhea. Unfortunately for some people the use of opium and its derivatives—heroin, Demerol, morphine, and codeine—is addictive, leading people to become dependent on their effects and abuse these substances. The plant-derived opiates are very similar in structure to some naturally produced neurotransmitters in the brain, the endorphins and encephalins. They mimic their effects. Another notorious substance is derived from the cocoa tree—cocaine, an addictive euphoriant (or mood elevator). Cocaine's main effect is to produce excitation by enhancing the release and inhibiting the recycling of the neurotransmitters norepinephrine and dopamine. We know that, even in very tiny doses, such substances alter arthropod behavior. For example, spiders exposed to a variety of psychoactive drugs produce bizarre web patterns. However, we don't know how these compounds prevent attacks on plants by predatory insects.

Some of the most unusual plant poisons are known to affect the nervous system of humans by producing hallucinations, a word that comes from the Latin *hallucinari,* meaning "to wander in the mind" or "to dream". Hallucinogenic plant toxins alter mental states, producing ecstasies, bizarre imagery, excitation, and sensory distortion. They are not mind expanding, as the late drug guru Timothy Leary suggested, but are toxins which in high doses can be lethal to humans. Among the more familiar hallucinogenic plants are certain species of morning glory that produce, among other compounds, LSD (lysergic acid diethylmide), which interferes with the neurotransmitter found in the brain called serotonin.

A number of mushrooms, such as those of the genus *Psilocybe,* which have been used in religious ceremonies for over 3,000 years, produce LSD-like substances. These mind-altering substances, psilocin and psilocybin, produce sensory alterations, kaleidoscopic visions, and all sorts of distortions described by the Beatles in their song "Lucy in the Sky with Diamonds" (LSD). The psilocybes are not the only mood-altering mushrooms: one, *Amanita muscaria* or fly agaric, a large, attractive, colorful toadstool, has been a favorite of artists in children's books, on bric-a-brac, and in cartoons as an umbrella for gnomes and fairies. The *A. muscaria,* with its colorful orange or yellow cap covered with creamy-white warts atop an 8-inch stalk, has a long-standing history of use by primitive peoples to induce visions. The Vikings used it for ceremonial intoxication, which sometimes led to murderous binges. The English word *berserk* derives its origin from the Viking clan that was notorious for such fungally induced rapes.

Fly agaric grows well in nitrogen-free, poor soil and is often surrounded by a circle of dead flies, killed by sipping the deadly juices of the cap. Some botanists have suggested that the mushrooms' poisons were not defensive weapons but offensive weapons that quickly killed the nitrogen-rich insects to provide fertilizer for themselves. As is often the case, fly agaric contains a number of toxic substances, including an atropine-like compound that produces fearlessness, excitement, and hallucinations, which is why it became popular. But its other toxins can be deadly; it is estimated that ten of these mushrooms can constitute a lethal dose.

One of the most potent neurotoxic groups of plants, members of the nefarious nightshade family, belong to the genus *Datura* which has a long history of use as a source of pleasure and also as a hallucinogen which often caused a stupor. In 1676, a group of British soldiers in Jamestown, Virginia, experienced a mass poisoning after ingesting the fruit of a species of *Datura*, which gained the popular name Jamestown weed, later corrupted to jimsonweed. This malignant genus contains a potent mix of toxins, of which scopalamine is the hallucinogen; but the other components, which are atropine-like if taken in large amounts, can cause seizures, coma, and even death. Another genus of *Datura*, the devil's trumpet, an ornamental species of shrub, has 12-inch-long, horn-shaped, white, musky-smelling flowers, whose beauty belies the plant's witches' brew of poisons that has caused deaths in people using the plant to make hallucinogenic teas. Other members of the *Datura* genus have frequently killed browsing herbivores.

The same poisons that are found in jimsonweed are also produced by black henbane, *Hyoscyamos niger*, which is also known as insane root and poison tobacco. This foul-smelling member of the nefarious nightshade family has a long history of medicinal use by humans, dating back to ancient Egypt. Its scientific name derives from the Greek, meaning "hog bean," a reference to its use in poisoning wild pigs. Its atropine-like effects have been responsible for a number of deaths in children. In fact, of the 400,000 cases of toxic plant ingestion reported annually by the U.S. Poison Control Centers, most of the victims are children. However, the primary impact of toxic wild plants is on browsing animals, causing losses of millions of dollars per year.

Plants' Diverse Arsenal of Chemical Weapons

Plants have evolved a huge armory of defensive chemicals that deter herbivores. Most of our knowledge about these so-called secondary compounds comes from studies on mammalian browsers and humans, since their impact has direct relevance to human well-being. We focus on the most potent and deadly of these chemical warfare agents. It is important to recognize that many of these poisons, in small amounts, are of great value as medicines, used to treat a variety of human diseases.

Irritants

Many plants prevent or slow predation by having irritant juices in their sap that immediately cause inflammation of the delicate tissues of the nose and mouth. Some of these scourges of the plant world are insidious villains that require a previous exposure to their toxins before having an effect. The toxins

Putting a Stop to the Devastation   Periodically the ravenous, leaf-eating gypsy moth caterpillar devastates vast forested areas. Its population explosion is eventually brought into check, as plants respond to its assault by manufacturing ever-increasing amounts of toxic defense chemicals.

are allergens that activate the mammalian immune system to produce antibodies, which interact with the toxin to produce inflammation, itchiness, and blistering. The most familiar of these are the so-called urishols, produced by poison ivy, poison oak, poison sumac, and their kin. It is estimated that between 40 and 75 percent of humans exposed to urishols develop an allergic sensitivity to them and then develop dermatitis, or inflammation of the skin.

Only one of the plants that cause dermatitis is truly dangerous, the manchineel tree, also known as poison guava. This handsome tree produces sweet, scented flowers and fruit called manchineel apples. Its milky latex contains several different toxins, some so potent that the Carib Indians used them as an arrow poison. In addition to skin irritants, the sap contains physostigmine, a potent nerve poison. Physostigmine deactivates the enzyme that normally breaks down the neurotransmitter acetylcholine, which then accumulates and has the same effect as nerve gases. This toxin is in high concentration in the fruit of the manicheel tree. In the eighteenth century, a garrison of British soldiers sent to the island of Saint Kitts ate the fruits, and many of them became violently ill or died.

Photosensitizing Plants

A number of widely distributed common plants such as buttercups, Queen Anne's lace, and Saint-John's-wort contain a curious toxin. This substance has no effect by itself, but once it is absorbed into surface tissues that are exposed to the ultraviolet rays in sunlight, it produces an exaggerated effect on the skin that is identical to severe sunburn. These compounds, called psoralens, absorb specific wavelengths of ultraviolet light and become energized. This energy is then transferred to oxygen, producing a highly reactive species of oxygen called a free radical, which can damage DNA. In areas of the skin not protected by dark pigments, severe blistering and sunburn will occur. One of the photosensitizing plants, Saint-John's-wort, whose active principle is called hypercium, has a long history of use as a folk medicine. Recently, it has been suggested that hypercium is a natural Prozac which can be used in the treatment of depression. Hypercium is widely marketed even though its efficacy as an antidepressant is poorly documented, and it is sold without any warning that taking hypercium combined with exposure to the sun can produce severe sun poisoning.

The Killer Toxins

Among the plants having an enormous impact on horses, sheep, cattle, and other browsing mammals are a number of truly dangerous plants whose seeds and juices are potentially lethal. Among them is the castor bean plant whose oil is widely used in industry as a high-quality lubricant and in medicine as a cathartic; however, the oil is free of the toxin ricin, which is concentrated in the seeds. Ricin is so poisonous that, if eaten, a single seed can kill a child, and three seeds a full-grown adult. During the cold war, ricin was used to assassinate a defector from the Eastern bloc. It seems that the clandestine operations of the KGB (and probably the CIA) had accumulated an armory of potent poisons to be used when circumstances permitted. Ricin seems to work its mayhem by causing a massive allergic reaction to normal proteins, resulting in an enormous breakdown of cells.

Cellular poisons are also produced by several species of mushrooms, particularly those of the genus *Amanita*. In the deadly amanitas are some of the most potent poisons known. Their common names attest to their lethality: destroying angel, death cap, and death's angel. Of these, *Amanita phalloides* (meaning "penis-shaped") contains more than twice the concentration of poisons of any other amanita. Ingesting a single mushroom can be fatal, and every year literally hundreds of human deaths are caused by eating amanitas. (Historically, the Roman emperor Claudius was murdered by his wife, who made a toxic soup from amanitas so that her son Nero would become heir to the throne.) The two toxins phallatoxin and amatoxin both kill cells, the latter's being an effective inhibitor of the enzyme RNA polymerese II, which is necessary for the manufacture of proteins. Once ingested, it is absorbed in the small intestine, where it creates cellular mayhem, and then is carried to the liver, where it devastates liver cells and becomes concentrated in the bile, which in turn is dumped into the small intestine, where it creates another wave of destruction. Amatoxin also causes extreme damage to the kidneys. The mortality rate is reported to be 30 to 50 percent.

Some plants also produce toxins that inhibit protein synthesis. The most nefarious of these is *Abus precatorius*, commonly called the rosary pea, crabs eye, prayer bean, or love bean. Its toxin, called abrin, is concentrated in its brilliantly red, shiny seeds, each with a large black spot at one end. A lethal dose of abrin is 1 part per

Chemical Warfare The destroying angel, *Amanita phalloides*, contains two cellular poisons so deadly that eating just one mushroom can kill a human.

1 million parts of water, making it one of the deadliest of the known lethal substances.

Some of the plants containing potent toxins are quite beautiful and are frequently grown as ornamentals in flower gardens. About 200 years ago, an English physician, W. Withering, found that extracts of the handsome flowering plant called foxglove, *Digitalis purpurea,* was highly effective in the treatment of some forms of heart disease. All parts of the plant contain steroid glycosides, the most potent of which is digitoxin, which increases the flow of calcium ions into the heart muscle. However, in high concentrations, the effects reduce function of the heart muscle and smooth muscle and can be lethal.

Another attractive but deadly flowering shrub that produces digitoxin is the oleander, *Nerium oleandrin,* whose pink and red blossoms make a showy display. This plant and its cousin, the yellow oleander, are known to contain over 50 different toxic compounds. Two of these compounds, oleandroside and nerioside, if ingested, can cause overstimulation of the heart muscle and death. Eating a single leaf could be lethal. There have even been reports of people using oleander twigs as skewers when barbecuing and, in this way, getting a lethal dose. But, because the plant sap also contains irritants that deter browsers, human poisoning is rare.

Historically, the most famous poisonous plant is poison hemlock, *Conium maculatum,* whose toxin, called conine, is a colorless volatile oil that has nicotine-like effects on the nervous system. In 399 B.C., the Greek philosopher Socrates, whose teachings ran counter to those in power, was forced to drink poison hemlock. The toxin is also found in cowbane, or water hemlock, and in fools' parsley.

The last of the truly deadly plants in this abbreviated summary is white snakeroot, which was the cause of milk sickness in the eighteenth and nineteenth centuries. Cows eating white snakeroot concentrated the ingested toxins in their milk, which not only made the calves sick but also caused humans drinking the milk to experience a variety of symptoms leading to tremors, delirium, coma, and death in about 10 to 25 percent of the victims.

Plant and Mushroom Bowel Irritants

In their toxic collection of defensive chemicals, most poisonous plants and mushrooms have compounds that irritate the lining of the digestive tract. When these reach the stomach, the inflammation there can cause cramps, spasms, and vomiting. (Almost all mammals find vomiting an unpleasant experience and quickly learn to make an association with whatever it was they ate that made them sick. Remembering the cause of their illness, they avoid that plant in the future. This type of learned response is called aversive conditioning.) Furthermore, some of the toxins, when absorbed, act on the vomiting, or emetic, center in the brain stem to trigger queasiness, nausea, and anorexia (loss of appetite) as well as vomiting. When they reach the large intestine, or colon, many of these same toxins produce an inflammation leading to diarrhea or dysentery which is usually preceded by severe cramping, bloating, and flatulence. In both vomiting and diarrhea, the body's responses are protective in that they remove the source of the illness. However, severe, protracted vomiting or diarrhea can be lethal by causing dehydration and disruptions in the acidity or alkalinity of the body. A number of bacteria and viruses also target the bowel and produce symptoms similar to those caused by plant toxins.

Plants that Prey on Animals

Plants have evolved a number of strategies for surviving in soils deficient in nitrogen and other nutrients. One tactic that has proven to be very successful is to establish an alliance, or symbiotic relationship, with another organism. For example, a great many plants establish a partnership with fungi attached to the outer surfaces of their roots. The fungi in turn send out thin filaments, which form mats and are highly efficient in sucking up phosphates, far more efficient than the root hairs of the plant. The phosphates enable the plant to thrive, and in return, the plant rewards the fungi with other nutrients. Another highly efficient symbiotic relationship is found in the legumes, such as beans and peas, that have incorporated specialized nitrogen-fixing bacteria in their tissues. These bacteria can take atmospheric nitrogen, the most abundant gas present in air, and incorporate it to form nitrogen-rich organic molecules which the plant needs. However, some species of plants evolved a less-giving strategy: they became predators that caught and killed animals, whose nitrogen-rich components were then digested and utilized by the plant. These plants have literally become carnivores, which is no mean feat, since plants lack not only the sensors common to animals but also muscles for movement and nervous systems capable of communication. They do, however, have other systems that do these jobs.

The most familiar carnivorous plant, the Venus's-flytrap, was described by Charles Darwin as "the most wonderful plant in the world." This plant is an active carnivore whose leaves act as snaptraps. Any insect landing on the open trap can activate hair cells that act as triggers, causing a slow-moving bioelectric wave, similar to that of a nerve impulse, to spread over the leaf and reach the springlike, thin-walled hinge cells along the leaf's midrib. The impulse causes these cells to become permeable and lose water, and the two halves of the leaf close. At the leaf's margins are a line of spines, so that when the leaf

A Botanical Minefield   The colorful flowers of the Venus's-flytrap are lined with fine hairs which, when disturbed, initiate an electric signal that causes the plant's two petals to close, trapping insects (such as the flies shown here) in a lethal embrace. The decaying insects provide the plant with much needed nutrients.

closes, they interdigitate, catching the insect in a cagelike trap. Other specialized cells on the leaf's surface secrete potent protein-digesting enzymes that
break down the victim's proteins and provide the plant with nitrogen. This
active hunting method is also used by several species of aquatic plants to feed
on crustacea.

At least 370 species of plants are known carnivores, capable of feeding on
insects, crustaceans, and even small rodents. One group, the aquatic bladder-
wort, is also an active hunter. Each stem has a multitude of weapons in its bulbous bladders, each of which, when deflated, is sealed tightly by a hinged,
waterproof door. At the margins of the bladder around the door are several finger-like projections covered with sensory hairs that can trigger the opening of
the door. When a prey animal activates the triggers, the door swings open
and water rushes into the bladder, carrying with it the victim. The valvelike
door then closes, and the tiny trapped victim is digested and its nutrients
are absorbed by the plant. There are more than 100 known species of
bladderworts.

Other carnivorous plants are passive hunters that lure their prey to them or
rely on chance encounters to capture prey. They employ a variety of weapons,
pitfalls, adhesive glues, and even so-called lobster pot traps which allow the
prey to enter but prevent its escape. As its name implies, the pitcher plant,
Nepenthes fusca, has a flask-shaped tube whose opening is attached to a prominent colorful flap and lips whose color and sweet aromatic secretions attract a
variety of small insects. Once inside the opening, the insects find themselves
sliding down the slippery, wax-covered, nearly vertical wall of the pitcher. The
bottom of the pitcher is filled with a vile-smelling brew of lethal enzymes that
digest the hapless victim. To add to the risk, the pitcher plant often attracts
crab spiders that build webs across the lips; those insects fortunate enough to
grab onto the web to stop their slide to doom soon find themselves terminated
and eaten by a spider.

The sundew plants have yet another weapon system: their sticky hair-covered leaves. Each hair is tipped with a glob of adhesive, which immobilizes its
prey. As adjoining hairs latch onto the struggling insect, the leaves are pulled
around the trapped prey and instantly start digesting it by secreting enzymes.

Chapter 8

The Ultimate Weapon System

The New Kids on the Block

Forty thousand years ago, a new species appeared on this planet and, in this micromoment of time on the evolutionary chronology, became the dominant species. Never in the course of evolution had a single species evolved to a state of total dominance over all other forms of life. The new kid on the block was the modern human, *Homo sapiens sapiens*, the wise wise man that mastered not only all other organisms but, to a large extent, most aspects of Earth's environment. This social being formed family groups, tribes, and nations which required concern for others and provided help to enhance the survival of their members. To succeed, such cultures required leaders, usually esteemed, trusted, concerned males. Humans used their evolving cultural capacities to redirect their innate aggressive programs into constructive displacement behaviors and created a system of rules that all members had to adhere to in order to maintain membership in the societal group.

The earliest true members of the genus *Homo* were curious and, being endowed with exceptional exploratory behavior, were migratory and constantly on the move. The currently accepted wisdom is that *Homo*'s origins were in East Africa and its members then spread to many parts of the world. In each region, *Homo* became the ultimate predator.

The Roots of the Ultimate Predator

Some 70,000,000 years ago, a new variant of the basic mammalian form appeared. Our ancient primate ancestor was a small, shrewlike creature that had a number of distinctive characteristics. These were a specialized, typically

The Evolution of Earth's Dominant Species   Our evolutionary origins may date back 530 million years to the Cambrian big bang, during which the first ancestral chordate appeared. We are vertebrates, and part of our ancestral tree (shown here) eventually led to the "new kids on the block"—modern humans. Although modern humans have been around only 400,000 years or so, they have become the top predators of all time and the dominant species.

primate brain, an increased use of visual and auditory communication, a decreased dependence on the sense of smell, a reduced number of young, a longer gestation period, a long adolescent period for maturation and social learning, and an increased life span.

The primates followed four lines of ascent from this primitive ancestor. The first, the premonkeys or prosimians, were small, tree-dwelling primates, long muzzled and long tailed. About 25 to 35 million years ago, the monkeys branched off the primate line, and finally, about 15 million years ago, the hominids (apes and humans) appeared.

The monkeys and hominids shared several new and important features. All had acute visual perception because of changes in both the brain and the arrangement of their eyes. Being blessed with stereoscopic, three-dimensional, color vision, they became much less reliant on the sense of smell, and via the sieve of natural selection, there was a decrease in snout length. Those parts of the brain used in smelling became less involved in chemical communication and more involved in emotion. The monkeys and hominoids also switched their sense of touch from long hairs around the snout to the hands. The hand developed a strong, grasping grip by which the fingers could wrap around an object. The fingers also changed, and claws were replaced by fingernails. These later primates showed a relative increase in skull size to accommodate the

The Thinker   Primates endowed with front-facing eyes, binocular stereoscopic vision, large brains, and a high degree of manual dexterity can not only learn by observation, but can also solve problems that require some degree of innovative abstract thought.

increased volume of the brain. Humans, then, are primates of relatively recent vintage whose evolutionary history has only partially been worked out.

From some early primate stem, the first humans arose. These were the hominids. This evolutionary development toward greater and greater complexity, leading to a group of advanced humanlike creatures, covers a span of some 25 million years. Humans did not derive directly from the apes but arose from the same common stock. This emergence of a common ancestor occurred during the Miocene period, about 15 to 25 million years ago.

The fossil record of hominid evolution is incomplete and rife with controversy, but by the Pliocene period an ape appeared in East Africa whose dental structures were much more humanlike than apelike. This creature, *Ramapithecus,* and its later relative, *Kenyapithecus,* may indeed be in the main line of hominid evolution, but until more complete fossils are found, we cannot be sure. What we do know is that there is a huge gap in our fossil record, maybe 10 or 12 million years, before we come to the period of human fossils, the Pleistocene period of geologic time.

Early Human Beings

The Pleistocene geologic epoch extended from about 3.8 million years ago until as recently as 10,000 years ago, when the modern period began. During this period of time, the pace of evolution suddenly became greatly accelerated. The reasons for the stepped-up evolution during the Pleistocene era include a series of dramatic and cyclical changes in climate that continuously demanded that existing organisms adapt. These changes were responsible for our evolution from the prehuman ancestral forms known as the australopithecines.

Two distinct populations of australopithecine hominids emerged. One population inhabited the plains of South and East Africa and was a rather small-boned, light-bodied creature. The other australopithecine forms were far more robust, larger, powerful creatures that inhabited more humid, forested environments in the same part of the world. The fossil remains of these creatures had limb-bone structure and dental structures that were more humanlike than apelike. Although there is evidence that they were tool users, their brains were still quite modest in size. While the fossil record is continually building, the current thinking is that these australopithecines gave rise to the early human forms of the genus *Homo.*

Homo

Olduvai Gorge in Tanzania, East Africa, is an archaeologist's dream come true, a treasure trove of fossil layers that provided Dr. L. S. B. Leakey and Mary, his wife, with a detailed view of life dating back almost 2 million years. Through years of meticulous, backbreaking work, the Leakeys probed their way through layer after layer. In 1960, they uncovered the remains of two distinctly human individuals only a few feet below the layer that had previously yielded the skull of a large australopithecine. The find included a jaw, parts of two hands, an almost complete foot, and a collection of primitive, human-made stone tools. The latter finding caused the Leakeys to call this new species *Homo habilis,* or "skillful man." At least 1.75 million years ago, this early being walked fully erect and had a brain case that was much larger than the australopithecine brain case. Some thought that Leakey's *Homo habilis* was really a highly advanced form of australopithecine, while others felt it was either an ancestor to or a member of the group known as *Homo erectus,* "erect man."

A number of almost complete and partial skeletons of *Homo erectus* have been found. The accuracy of the dating of these remains is still questioned, but they may go back to well over 3.7 million years. The skulls of these creatures not only showed a bone structure similar to that of modern humans; it also provided a brain case ranging from 750 to 1,200 cubic centimeters. These early humans had the beginnings of an organized culture. They made tools, observed rituals, and used fire. The best of the skulls of *Homo erectus* was also discovered by Dr. Leakey at Olduvai, and reliable dating evidence shows that these creatures were still alive only half a million years ago. More recent findings from northern Kenya by Leakey's son, Richard, and other findings in Ethiopia suggest that *Homo erectus* was an inhabitant of East Africa at least 3.75 million years ago. Given these data, it is apparent that both the australopithecines and *Homo erectus* coexisted and probably competed well into the middle of the Pleistocene period. Then the australopithecines became extinct.

Homo erectus was also destined to disappear, but its progeny were to diversify and give rise to two groups of true humans, *Homo sapiens neanderthalensis* (the Neanderthal) and *Homo sapiens sapiens,* the modern form. Today there is controversy about the classification of the Neanderthals, and some think them to belong in a separate category, *Homo neanderthalensis.*

Neanderthals

The Neanderthals derive their name from a skeleton found at Neander in Germany. Other Neanderthal skeletons have been found at many other sites in Europe and the Middle East. These Neanderthals had large, flat faces, a broad, flared nose, and short, very muscular, powerful, thickset bodies. They could walk and run as well as any short-legged modern human. (Early ideas of the Neanderthal's being hunched over were based on the remains of a Neanderthal crippled by arthritis.)

The Neanderthal brain was comparable in size to the brain of modern humans. Artistic, cave-dwelling hunters of considerable skill, they left us ample evidence of their culture from Russia to the English Channel to the Middle East and North Africa. They made crude stone tools and used 8-foot-long lances of sharpened, fire-hardened wood, but apparently never devised thrown weapons. Why these seemingly successful creatures suddenly vanished is not clear. They may have been unable to compete with the more-advanced humans; they may have interbred with modern humans and assumed their culture and their collection of genes. But, for whatever reason, about 30,000 years ago, *Homo sapiens neanderthalensis* disappeared, leaving only *Homo sapiens sapiens* as the end product of hominid evolution.

Sapiens Sapiens—Indeed Wise

As humans evolved, the size of their brains increased. The large, efficient brain and several other new traits, such as freedom of the limbs, keen color vision, manual dexterity, and stamina, allowed humans to become the dominant species on this planet. One factor that allowed their large brains to accumulate the knowledge they needed to reach such a dominant position was a long life span, for humans have a longevity shared by few species. Thanks to this long life span, humans had the time and the mental equipment to learn the habits of their prey and where they were most vulnerable. They learned to use insight gained from observation, to make snares and traps, to hunt cooperatively, and to consistently improve their skills. While they had neither fangs nor claws, their large brains and hands allowed them to create their own fangs and claws.

Humans first used available objects as weapons and later made them of stone, bone, fire-hardened wood, and volcanic glass. Early humans were prodigious hunters, killing animals many times their size and doing it with cunning and guile. They extended the range of their weapons, first by throwing stones and spears and later by using slings, bows and arrows, and bolos. They increased their choices of foods; they domesticated animals and learned to grow their own foods. They learned to dig wells and to store and preserve food for those times when food was scarce.

Finally, humans developed systems of culture that included art, music, and religious ritual. No other species has ever gone into such a phase of secondary evolution. What was so different in the human brain that permitted such changes to take place? After all, the human brain is basically similar to that of all mammals and even more similar to the brain of all primates. Was it size alone, or did new specializations occur?

Among the animals with large brains (including other primates, whales, and dolphins), humans alone have developed different functions for each half of the thinking brain. In all other mammals, both cerebral hemispheres perform the same tasks, but in humans there is a marked separation of function in which one half of the brain becomes dominant and is able to express itself verbally. So, in a sense, we humans have two brains. In most human adults, the left cerebral hemisphere controls verbal expression (speech), analytical abstract processes, and the precise movements of the right hand and dominates the mute but artistic right cerebral hemisphere. The language dominance of the left hemisphere was recognized in the early 1800s when it was discovered that damage to the left hemisphere produced a variety of speech defects, depending on where the brain damage occurred.

Right- or left-handedness appears to be genetically controlled (about 6 to 11 percent of the population of the United States is left handed). In left-handed people, the degree of left-hemisphere dominance is not clearly established, although in most left-handers the speech centers are still in the left hemisphere. While this is generally true, there are exceptions, and it has been estimated that about 15 percent of left-handed people and 2 percent of right-handers have speech centers in both hemispheres of the brain. Early in life, the capacity to develop speech centers in both sides is equal, but perhaps as early as the age of one year, dominance of the left hemisphere begins to manifest itself. At about one year of age infants learn to perform two-handed tasks using two distinct grips; a strong, stabilizing power or holding grip (usually with the left hand) and a precise, manipulating grip, usually with the right hand. Monkeys and apes also have power grips and precision grips, but they use right and left interchangeably.

There is good evidence that the degree of left-brain dominance is in part culturally determined, and some experimental studies have shown a lessened degree of left-brain dominance among children reared in an environment where the use of an extensive vocabulary is not important. However, these same children outscored their more verbal neighbors in pattern- and form-recognition tests. What does the right hemisphere do? Evidence is accumulating that strongly suggests that the right half of the brain, although incapable of verbalization, is artistic, rhythmical, musical, and emotional.

In normal humans, the right and left halves of the brain are connected by a giant communication channel called the corpus callosum, a white bundle of nerve fibers about 3½-inches long and ¼-inch thick containing about 400 million nerve fibers. We now know that there is an enormous flow of information between left and right hemispheres, and any sensory input to one half of the brain becomes available to the other half within a fraction of a second. Furthermore, it appears that the same bit of learning is stored in both hemispheres. What probably happens in normal functioning is that we alternate using our two brain halves.

In a verbal or quantitative task, the electric activity of the dominant left hemisphere shows an awake and active pattern of response, while the minor hemisphere, the right, shows a pattern of brain waves similar to that seen in a brain during light sleep. However, the opposite occurs when subjects are asked to perform artistic tasks or pattern recognition or to remember musical tones. In this case, the nonverbal right hemisphere shows the alert, awake pattern of brain waves and the left half "goes to sleep." From an evolutionary point of view, what does this unique development of two separate but mutually cooperative brains in the same head mean? Certainly, the distinct separation of motor control into power and precision allowed us to become toolmakers. The evolution of discrete speech centers in the dominant hemisphere allowed us to talk.

The Gift of Speech

Of all animals, humans have evolved the most efficient, versatile, and rapid system of communication. In the process of our development, the sense of smell and chemical communication became less important and the use of visual nonverbal signals became more subtle and complex. But the greatest advances were in the development of an extremely complex system of vocal communication. No other animal even approaches the repertoire of sounds and skills that humans use to communicate. These new skills obviously required structural changes in both the vocal apparatus and the speech control centers of the brain. But long before humans used vocalization as their main form of communication, they relied on a variety of nonverbal signals.

Wordless Communication

The smile, the clenched fist, the nod, the frown, the shrug, the blush, the kiss, the touch, the gesture—all are signals of communication. These signals are culturally learned (although some experimental evidence with apes suggests that the response to some visual signals such as the facial pattern of rage may be inherited).

Unquestionably, early humans, with their versatile hands, used sign language along with their still-undeveloped vocal apparatus. Among primitive tribes such as the African Bushmen, sign language is still widely used during the hunt, when sound could scare off the prey.

The Vocal Apparatus

Apes have the same vocal apparatus as humans. They have lips, a tongue, a pharynx, and a larynx with vocal cords, but they can't speak. How did our vocal equipment change to permit the gift of speech? The tongue and lips became larger, thicker, and more versatile. These structures were used to shape sounds into words. At the same time, the larynx moved further down the throat, and the pharynx, which makes the vowel sounds ah, ee, and oo which are basic to all languages, became significantly larger. The base of the tongue became anchored to the larynx. Now the basic equipment for speech was ready.

The sounds made by moving air over the vocal cords are altered by changing the shape of the sound chamber, the pharynx. Language, then, is produced by providing a continuous and fine control over the muscles of the pharynx. Since speech requires an exquisite coordination of the fine muscles of the tongue, lips, pharynx, larynx, and respiratory system, it is necessary to examine those specialized control centers in the brain that allowed humans to become the verbal beings they are.

The Brain and Speech

It is thought that the small mutations necessary to rewire the brain for speech took place in the last 2 million years and, in all probability, only as recently as a few hundred thousand years ago. Certainly, the fossil record shows an increasing brain size: the modern human brain is 100 percent larger than our australopithecine ancestor's and 65 percent larger than the brain of *Homo erectus,* both of whom still existed about 2 million years ago. With the increase in brain size, there evolved a series of speech centers usually located on the left side of the brain. There are two major speech centers, interconnected by a large number of nerve fibers, which carry the informational traffic needed for speech. The speech control centers consist of an *integrating center,* which takes the signals, usually sounds, that prompt speech and relays them to the *language selector,* which retrieves relevant memories to help select comprehensible language. Damage to the language selector region may result in jumbled, meaningless speech. From the language selector region, signals are relayed to the *speech control center,* which sends signals to the lips, tongue, pharynx, voice box, and facial muscles. Damage to this area produces a very slow and impaired speech. Needless to say, nonverbal signals and verbal signals are synchronized and combined to enhance the efficiency of the communication system.

After the development of a verbal language skill, came the development of a written language skill. This skill has existed for only 10,000 or 15,000 years, although symbolism as seen in cave paintings goes back to considerably earlier times. These paintings, however, cannot be equated with a written language. This brings us to another one of our unique gifts: the remarkable precision grip of the human hand. This precise, manipulative skill allowed humans to become the toolmakers, artists, and musicians that they are today.

The Human Hand and Freedom of the Upper Limb

Several hundred million years ago, all vertebrates developed a basic plan for the forelimb. This structure evolved as a pentadactyl limb (five digits, or fingers and toes) and, in the process of evolution, was modified to meet the needs of each organism's way of life. Humans display this basic pattern with several significant changes. These remarkable changes in the hand were the product of our erect, two-legged mode of movement; in other words the lower limbs were specialized for propulsion, giving a degree of freedom to the upper limbs that no other bipedal (two-legged) animal had ever had. This upper limb has a fore-arm capable of rotation and a hand with an enormously increased range of movement. The hand had the ability to grasp strongly with the thrust of pressure toward the side of the little finger. At the same time, there developed the opposable thumb, capable of making good contact with the forefinger (as in holding a pencil) and with all other fingers. So, the human organism developed three grips: a power grip, a hook grip, and a precision grip. This delicate precision grip that allows us to write and to manipulate objects with great dexterity is usually a gift given to the right hand and controlled by the left brain.

The Bipedal Gait

Walking upright with the arms swinging is peculiar to us alone. This efficient kind of movement makes excellent use of gravity, conserving energy and providing remarkable endurance as well. In the process of evolving an upright posture and bipedal gait, humans acquired a number of variations on the basic Primate theme. The bones making up the pelvic girdle (the hips) adjusted to the erect stance and provided a wide range of hip movement. Though flexible and free, the kneecap made little allowance for rotary motion. The proportions of the leg bones lengthened, allowing a longer stride. Walking and running, complex movements requiring balance, muscle power, and complex control of the sensory and motor process, involved some significant changes in the foot. Unlike that of the ape, which uses its big toe to grasp objects and walks on the side of its foot, the human foot became arched into an arrangement of flexible, springy levers. The big toe lost its grasping ability but acquired a strong pushing action.

With these skeletal changes the calf muscles proportionally increased in length and size, along with the powerful thigh and buttocks musculature. The large buttock muscles are particularly important because they pull the body forward and over the leg with each step. Apes, with their reduced buttock musculature, cannot do this; consequently, they don't stride forward but shuffle instead. Another advantage seen in the human leg is the ability to lock the leg at the knee, forming an erect, rigid pillar without having to waste energy by contracting the thigh muscles. Although humans were capable of considerably greater speed than many animals, it was not speed but stamina that enabled them to become such efficient predators. Many of the animals that could outrun humans simply didn't have the stamina to escape from their tireless pursuit.

Structural Adaptation to Hunting

Humans were large enough to scare off many potential predators, yet not so large that most of their time had to be spent eating. Since they were tall and upright, they had the advantage of being able to observe distant objects from their own natural observation tower. At the same time, since they could look down, they could see prey that shorter predators would miss. Their three-dimensional color vision allowed them to perceive form as well as color patterns, so that the protective coloration of many animals was of no avail. Part of human stamina was related to a capacity to tolerate heat stress better than many animals. (Humans have more sweat glands than any other primate and are able to use evaporational cooling during strenuous physical exertion.) Furthermore, humans were not burdened by a heavy coat of hair; having short, fine hair, they were indeed naked apes. This was an obvious advantage in hot regions but a detriment in cool or cold regions. They could counter the weather, however, by using animal furs, shelters, and fires.

Assaulting the Wall

Trying to understand the inner workings of the brain is a daunting task, and nature has thrown up an array of barriers that limit our efforts to gain insight into the seat of behavior. It's true that an army of researchers, fortified with a variety of highly advanced technologies, is relentlessly assaulting these barriers to gain some understanding of how the brain works, but as the millennium approached, much of what the brain does and how it does it remains a mystery. As a scientist, I know that one basic premise of the natural world is that the unknown is knowable; so, the search for insight will continue into this, the most complex of organs.

Why is it so difficult to obtain insight into brain function? Part of the answer lies in the very complexity of the brain. It begins with developmental processes dictated by a genetic program, and even in the uterus the expression of those genes is modified by an ever-changing group of variables. Nerve cells are born, some die in a programmed neuronocide, and others migrate and make connections. The process goes on even after birth, but before an infant is a year old, the capacity to grow new nerve cells ceases, with the possible exception of the neurons involved in the sense of smell. Nerve cells do retain the capacity to repair damage to their parts, and the ability to create new connections, or circuits, continues throughout the life of a healthy individual; the brain is also shaped by the person's total life experience. If deprived of specific nutrients during the early years of life, the brain can't undo the damage even if it is provided sufficient nutrients forever afterward. The brain can also be altered by a variety of infectious diseases, particularly some viruses that are programmed to attack nerve cells.

But, the main factor shaping the brain is experience. Some experiences can only have an impact during a very limited time, the critical period, during which certain kinds of relatively permanent learning occurs. Later, the brain can acquire these capabilities or undo that learning only with great effort. The brain is a learning machine that can store and retrieve information (or memories) on demand. Sometimes, however, the retrieval mechanism is actively suppressed or interfered with, and then we can't access our memory banks. Learning involves, of course, the sensory input systems and requires inputs from both the attentive systems and motivational systems. Nevertheless, after a lifetime of research on memory and learning, Karl Lashley concluded that right now all we know about memory is that it exists. Some of these memories that are formed, stored, and relatively hardwired into animals' brains are the products of early play.

The modern human brain attained its current shape and size only 50,000 to 100,000 years ago, and all humans since then have been equipped with a mental tool box to guide the hands that created the tools that allowed humans to become masters of the world. Among the ingenious devices these brains created were a vast variety of weapons. Weapons, in one form or another, appear in the fossil record long before the modern brain took its current form. Hence, weapon-building for the relatively puny, fangless, clawless, early humans was a major selection factor in the evolution of the modern brain. Given a versatile apparatus that has the equipment to learn from experience, observation and imitation, and has an intuitive capacity to understand basic physical principles, it is not surprising that even young children can create an amazing armory of weapons.

As children, we learned that a tube-like soda straw could act as a blowgun, providing we had a projectile, a spitball, and a propulsion force, a forcibly exhaled breath. We learned that elastic rubber bands, when stretched, stored energy, as did wood and any variety of flexible materials. The stored energy, when released, could be used to propel any number of objects. We made slingshots, bows and catapults that fired a variety of missiles at distant objects. We intuitively understood that wind and gravity effected the path of our launched projectiles, and hence, could calculate trajectories needed to hit the target. We always sought to build better launchers and attain better accuracy.

In a different time, when I was a child and firecrackers were not illegal, I built a toy cannon that used the explosive force of a large firecracker to propel a steel ball bearing with enough force to go through a tin can. Years later, when I read the history of the genius Santiago Ramon y Cajal, I discovered that the father of modern of neurobiology, and winner of the first Nobel Prize in Biology and Medicine, also had built a cannon as a child. His target was a local church, whose rector didn't take kindly to having the building's masonry shattered, and young Cajal was severely punished.

I related this story to my physiology class of 120 biomedical engineers and asked how many of them had ever built a bow, a slingshot or cannon, and to my surprise, over 80 percent of the class held up their hands, and 50 percent of the students were women. It seems that the human toolbox (the brain) that creates weapons has an inherent propensity to do so. This was recognized by the founding fathers of the United States when they incorporated in our constitution "the right to bear arms," a right that is exercised much too often and is too often abused.

War Games

One characteristic of evolution is that pervasive structures and/or behaviors are purposeful, are in part innate, and have functional significance. Certainly, both cooperative play and solo play provide young animals with the tools they will need later in life in order to be part of a group, to hunt, to cooperate, to compete for status within a social hierarchy, to adjust to a varied and often changing environment, to mate, and to rear their young.

As one goes higher on the evolutionary scale, the animals with bigger brains develop the most ornate and complex play patterns. This is particularly true in gregarious animals (dolphins, killer whales, wolves, lions, baboons, chimpanzees, and humans), where mutual cooperation within the group is essential to the survival of the individual.

Animals' brains are innately programmed to produce play, in the form of exploratory and manipulative behaviors. Young animals always manipulate objects for no other reward than the manipulation process itself, and when normal young animals are placed in a novel environment, they always explore it, tentatively at first and then more vigorously as they learn more about it. This drive to explore (also called curiosity) is so strong that animals will work at it with great motivation and will even forgo food and tolerate pain to satiate it. If they are prevented from exploring, the subsequent development of their brains is impaired.

This was shown in an elegant series of experiments comparing the brains of rats reared in caged isolation with those of rats reared in a stimulating environment. It was clear from the results that the stimuli provided during development had significant beneficial effects on both brain chemistry and brain structure. Other studies showed that young rats reared by mothers that lacked curiosity also showed little exploratory behavior and that young monkeys deprived of the opportunity to interact with other young monkeys during a critical time in their development never developed the skills needed to socialize as adults and were even deficient in rearing their own young. While there are no adequately controlled studies on humans, there is an abundance of clinical observations that suggest that critical periods for learning socialization do exist and that people deprived of those early experiences often grow up to be dysfunctional adults.

Watching young social animals and young humans at play, we see the same scenarios repeated over and over again. However, patterns of play are often altered in context and sequence as the roles of the pursuer and the pursued change again and again. Observing animal and human play, one is struck by the creativity of these inexperienced players. They innovate, improvise, and cheat by bringing in new strategies to their games. But most of all, they learn to compete.

Playing not only is a cooperative phenomenon among the young of the same species but also can be and is a solo exercise which involves lots of repetitive, energetic gymnastics and manipulation of nearby objects. Watching the pervasiveness of play and games in young animals, both predators and prey, leads to only one conclusion: play is preparation for survival battles to come.

Although the games may be rough and tumble, they seldom result in serious damage to the contestants. One notable exception is play among hyenas, where dominance and belligerence is the domain of the females. If the female gives birth to twin females, the two cubs fight to the death, as we previously noted. While this seems counterproductive to the need to increase the species, it does select for aggressive traits in the future leader of the clan. But what exactly is aggressive?

On Aggression

The term *aggression* carries considerable baggage in sociological and psychological literature; usually, the focus is on socially disruptive and/or pathological behavior. Consequently, "aggressive behavior" has become synonymous with violence and destructiveness, and "to be aggressive" has become pejorative. That it does carry such a negative connotation is reflected in the increasing use of the term *assertive* in describing human behavior.

The simple point here is that the term *aggression* is deeply entrenched in scientific literature as well as in general usage and that it has become value laden. A definition for our purposes here must be in biological terms which are unambiguous across the different disciplines and free of disciplinary biases. To arrive at such a definition of *aggression,* we must unburden the term; we must shear away the layers of human bias and preconceptions which surround it, whether semantic, disciplinary, social, or historical in origin.

Aggression refers to a set of behaviors which is exhibited by organisms at all levels of evolutionary development. These are adaptive behaviors that resulted from and still serve the survival of the species and the individual. Given these assumptions, any definition of *aggression* must include a broad spectrum of behaviors, a spectrum which will find its minimal expression in lower organisms and its maximal expression in higher animals that possess a fuller repertoire of behaviors, both more subtle and more complex than those exhibited by more primitive organisms. Aggression is a spectrum of social behaviors with a competitive dynamic between members of the same species. Since competition is the driving force behind evolution, it is to be expected that it has an innate or inherited program. Furthermore, given that all mammals share a large number of the same genes, animal studies provide some insights into human behavior. So, the primary cause for human aggression is the evolution of behaviors that are advantageous for survival and selected for over time.

Of course, genes don't act in a vacuum, and the expression of the innate program depends on the total environmental history of the organism. However, the potential for aggressive behaviors must be evoked by some appropriate situational stimulus, and herein lies the final determinant. There is a constraint or series of constraints which brings to play learning what society (the collective wisdom of the cooperating group) allows as acceptable expressions of aggression. All human societies have rules that govern behavior; sometimes they are incorporated into belief systems (religions) or political systems (families, tribes, clans, or nations). For the individual who violates the rules, there is punishment which is prescribed by law or by expulsion from the society.

However, the system has some serious flaws. At times, notwithstanding any societal constraints, individuals violate the acceptable rules and commit behaviors that are at the far right of the spectrum. They kill, maim, and abuse other organisms of the same species, often for no apparent reason. There are also episodes in which some members of a given society kill members of their own society who disagree with their views of life. And there are wars waged between two societies of humans which have brought about massive death and destruction. Why have these apparently senseless, counterproductive mass behaviors occurred with such frequency in the course of human history? Given what we know about other animals, it might be more sensible to ask why these behaviors have not taken over and done us in.

The dominant left brain is an amazing survival machine that analyzes, thinks, and makes risk-benefit and cost-benefit decisions related to survival. It invents and produces weapons for defense, some of which are weapons of mass destruction (nuclear weapons). Our left brain having done so, we have, for the past fifty years, deterred major conflicts by having the capacity to mutually assure destruction of each other. Deterrence works because the left brain sees a lose-lose situation and opts not to play.

Fear and Anger

The brain assesses any sensory input that conveys threat and sends us into one of two responses: fear or anger. Both situations result in the activation of the body's emergency response systems to provide behavior appropriate to survival. These behaviors are called coping responses and involve similar mechanisms that mobilize all the physiological assets needed in high-stress situations. The left brain is put on high alert, calling on its accumulated knowledge and thinking capacity to make wise risk-benefit decisions, namely, to stay and meet the threat (fight) or to retreat (escape). The fight decision is reasonable when the forthcoming confrontation is deemed winnable on the basis of previous experience. The escape option usually is chosen when the threat is something new and the person has not yet developed strategies and tactics to adequately cope with it. The basic idea is that "he who fights and runs away lives to fight another day." With either option in an emergency the sensory, motor, and analytical systems must be brought to a state of maximum alert by the brain's attentive system.

The attentive system consists of a dense collection of nerve cells in the brain stem. This network selectively adjusts itself to sort out incoming relevant information from irrelevant information, to suppress information that is not relevant, and to activate the appropriate parts of the central nervous system to the call to battle stations. The anatomy of the eye, the inner ear, and the receptors in skeletal muscle not only respond to the environment by sending input signals to the brain, but they also receive signals from the brain which can selectively increase or decrease the sensitivity of the sensory receptors in such a way that relevant inputs are amplified and irrelevant inputs are damped down.

The third component of the call to battle stations is to bring the data-processing circuits of the brain to a state of generalized alert and then to focus attention on the pending threat. At the same time, even prior to responding to the threat, the brain activates the hormonal components of the body's emergency system, the adrenal glands.

Emergency

The paired adrenal glands, located just above the kidneys, consist of an outer and inner component. The inner component, under direct control of the nervous system, can release two different emergency signals (epinephrine and norepinephrine), which act in differing degrees on many different organs. Unlike the short-acting substances released at nerve endings, these messengers are carried by the bloodstream to all body parts and exert their effect for minutes and even hours. The multitude of responses are all purposeful in that they prepare the body to respond to the emergency: the strength and rate of contraction of the heart increases; the blood pressure rises; and the blood flow to the skeletal muscles, the respiratory system, the heart, and the brain increases. Emergency responses require lots of available energy, and to this end the adrenal hormones convert stored sugar from the liver and muscles (glycogen) into readily available blood sugar. At the same time, the blood flow to the bowels, sex organs, kidneys, and skin decreases to a trickle, just enough to minimally maintain those organs. In both fear and anger, the blood flow to the fingers decreases, so that the fingertip temperatures drop as much as 10°F. Saliva flow decreases and the mouth becomes dry. The diameter of the pupils of the eye changes, and the blood vessels of the face either constrict (close) or dilate (open), producing the well-recognized pallor of fear or flush of anger. These changes are a visible signal to the opponent.

Emergency

Emotion poured from me in waves of flood
And fire, converting sentient flesh and bone
To steel as cold as stone, the coursing blood
Congealed. Machine controlled by brain alone.
My body, mechanized and energized,
My limbs moved swift and sure, correct
As gears in every turn and touch, devised
To reconstruct, to overcome defect.
And with the crisis passed, when I had done
The best that could be done, the flesh of me
Returned to flesh, the blood to heat of sun,
The pulse, the breath, the strength to normalcy.
Emotion, too, reclaimed its former role
And left me quivering beyond control.

SAM BERGMAN

The other part of the adrenal gland, the outer portion or cortex, is turned on when the command center of the emotional brain (the hypothalamus) signals the pituitary gland to turn on the adrenal, which produces another group of emergency hormones that have a multitude of effects, including a further increase in blood sugar and increased excitability of nerve and muscle. However, prolonged exposure to stress can deplete the stores of these hormones, producing a phenomenon known as combat fatigue. (During the Korean War, it was found that troops that had been in combat for protracted periods lost their desire and ability to continue the fight because they had used up the emergency hormones in the adrenal cortices.) Further preparations for the coming fight are made by the emotional brain, or limbic system.

The evolutionarily ancient emotional brain is composed of a variety of components linked together to produce motivational and emotional responses, some of which we considered earlier in the book. These responses involve both the external and internal, or visceral, motor systems. Commands sent to the skeletal muscles cause facial muscles to contract and put on the so-called game face. The jaws clamp shut, the eyes narrow, and the lips are drawn taught. The whole set of the face is changed, sending a highly visible signal saying, "I'm ready; don't mess with me." The body language also changes: the fists clench, the muscles increase their tone, and the whole body posture signals, "Do not get too close." You can see this attack set in sporting contests when you look at the players about to begin a match. I could see it during the war in my crew members; but I could also see the face of fear in one of my crew, who was so terrified about going out on a mission that he shot himself.

Fear is a protective response to threat. Like rage, it is seated in the limbic system, particularly in the amygdala. Fear is learned and can be overwhelming, or it can be overcome. It is essential for survival. All of us at times are afraid, usually for due cause. The same soldier who wins a medal for gallantry in action may, on another occasion, become immobilized by fear or run away. Like rage and anger responses, fear has a strong visceral component. People get the shakes, they may vomit, they certainly will sweat profusely, and they may lose control of their bladder or bowel. One old sergeant was heard to tell his green troops going into combat for the first time, "Today most of you will need your helmet on your bottom, not on your head." Fear is an essential survival behavior, and a fearless soldier is usually a dead soldier.

Having mobilized all the body parts, the brain goes into an attack set in response to commands issued from the hypothalamus, the final common pathway for signals from the limbic system. But the attack set must be activated by other signals from the thinking brain which send the "go" signal to charge. Herein lies the dilemma: the thinking brain weighs the options, assesses previous experiences and the immediate situation, and then makes the decision to go or not to go.

For example, studies of mouse-killing behavior in rats show that individual rats "normally inclined to kill mice were more likely to kill during hypothalamic stimulation than non-killers." Thus, the electrically elicited response was probably not determined by specific functions of the tissue under the electrode

Warning Signs Humans, like most mammals, look at the faces of beings they encounter, most of the time focusing on the eyes and mouths of those faces. The communication of emotional states such as anger is conveyed nonverbally as body language. The angry face shown here is a warning sign of a potential attack and a signal to "back off."

but by the personality of the rat. The second factor to be considered is the role of a particular animal's previous experience. For example, in studying the mouse-killing behavior of wild Norway rats, it has been found that even these killer rats will not attack mice if they are familiar with them. Even brain lesions, which routinely initiate aggression in wild rats, do not induce mouse-killing behavior in rats that have had previous experience with mice prior to the operation. One could conclude that a soldier is less inclined to kill an enemy soldier if he comes to know him.

The most-perplexing examples of inhumanity find their expression when it comes to belief systems where anyone not adhering to a particular belief system is relegated to nonhuman status and thus can be killed without conscience or remorse. One example of this state of affairs is the Spanish Inquisition. What is even more perplexing is that sometimes humans are not content to simply kill their opponents but indulge in the most insidious forms of torture that impose maximum suffering before death.

Humans have a long history of making spectacles of suffering and death. Ancient societies built great arenas to stage battles to the death. During the French Revolution, huge crowds cheered the beheading of the nobles who had exploited them. Indeed public executions have always been well attended by enthusiastic onlookers, and these still occur today. Our capacity to enjoy violence extends to many of the so-called contact sports, from the charade practiced by professional wrestlers to boxing, football, rugby, hockey, and even basketball. The crowds oooh and ahhh with each impact. These contests appeal to the pleasure centers in the brain that appease the appetite centers for violence. Why?

Why?

Just as the centers for appetite (for food, sex, and drink) are satisfied by some sort of biochemical satiety signal, so, too, are the centers for unabated violent aggression satiated either by participating in the payback or by experiencing it vicariously, observing and even applauding it.

The thinking brain can learn compassion and love. It can create works of beauty and wonder. It can be charitable, kind, considerate, and caring—all truly noble characteristics. However, the circuits that produce these characteristics are plastic; that is, they can be modified by rationalization. The mind can justify anything by mental gymnastics, particularly if the targeted enemy has behaved in a manner that violates its precepts of what is right. While laws and religion forbid killing, they paradoxically also approve of it. The ancient call of "an eye for an eye and a tooth for a tooth," getting even, or vengeance, is pervasive in human societies. Killing in self-defense is universally accepted as a justifiable response to a threat to one's own survival. The neural machinery for killing is there. We all have a potential dark side which has been exploited by charismatic despots, leaders with the capability to manipulate the masses to target other peoples, to reclassify those other people as inhuman or as threats to the institution or state. All that has to be done is to bypass previous learning. The brain can be reprogrammed, not only to destroy the threat but also apparently to enjoy doing so. Such pleasure is part of the brain's reward system.

The late, great football coach Vince Lombardi said, "Winning isn't everything; it's the only thing." When a person deals with either a real competitive threat or an imagined one and wins, the brain rewards the victor with a squirt of chemicals that produces euphoria, a feeling so good that it's almost ecstasy. "I won; I'm still alive." These euphoriants are addictive as has been demonstrated in controlled experiments and as has been attested to by many anecdoctal observations. I spent several years working with people associated with the violence center at Massachusetts General Hospital where Frank Ervin and Vernon Marks were studying people with episodic violence. These were intelligent, educated people who were sometimes provoked to commit violent acts. All of the patients who committed these acts somehow viewed them as pleasurable; they would smile, relax, and even drift off to sleep, satiated by their act. It was a small test group; studies with implanted electrodes revealed abnormalities of the amygdala. Partial removal of the amygdalae caused more than half the patients to lose their urge for violence. Marks and Ervin's book, *Brain and Violence,* evoked a storm of opposition, particularly from analytically oriented psychiatrists, and the procedure was suspended. There is no argument that the brain is the seat of violence, but the critics felt that we still don't know enough about its workings to go in and snip out various pieces.

The addictive nature of the competition involving survival can be attested to by the behavior of those who have been in life-threatening situations and survived, yet volunteer to go back again and again. During the war, I was placed in a replacement detachment to provide bomber crews. While some of us were naive and inexperienced, there was also a large number of seasoned pilots who had flown enough missions to be excused from further combat. We called them "war lovers." They loved the thrill of surviving; they were addicted to the brain's spurt of euphoriants. These veterans were all men, which raises a related question: Is there something distinctive about men's brains and body chemistry that has predisposed them to violence? Are men really from Mars (the god of war)?

Male Hormones and Behavior

Does the presence of testosterone during embryonic and fetal development in the first few weeks of life affect brain development? We know that by the middle of the first trimester of pregnancy, the developing testes start to produce increasing amounts of testosterone. By the middle of the second trimester, hormone production reaches a peak; it declines in the third trimester, and then shortly after birth, there is another brief increase before the levels fall off to practically nothing. Thus, during the period of brain development in males, testosterone is present. Some investigators have reported that in males some populations of neurons are depleted during this time. The question is, are there differences between the brains of males and females, and if there are, how do these differences express themselves in behavior?

In the past decade or so there have been a number of studies purporting differences in brain structure between males and females. The reported differences are subtle and not well substantiated, but there are enough studies to support the contention that there are differences in male and female brains. The problem is, as is the case with any cause-effect behavioral studies, that there is not clear-cut evidence of brain-related differences in behavior.

There is a large number of laboratory studies in a variety of mammals linking the male hormone, testosterone, to assertive behaviors, including assaultive behaviors. However, almost all of the research, when viewed critically, is somehow flawed and difficult to relate to human behavior. Nevertheless, the pervasiveness of the suggestive evidence has led several states to enact castration laws, most punishing habitually assaultive violent men with the removal of the source of most of their male hormones. In addition, there have been attempts to treat chronically violent men with drugs that inhibit testosterone, but the results of studies related to this approach are also subject to criticism.

Cross-cultural studies in humans show a general universality of male violence directed at other men and women of their own community. When males act violently toward others in the communities, the violence escalates to include infanticide. History has shown that, in their quest for dominance, males try to win no matter the cost. In a recent book, *Demonic Males: Apes and the Origin of Human Violence,* Richard Wrangham and Dale Peterson attempted to explain the dark side of human nature by using the chimpanzees, *Pan troglodytes,* of the Gombe Forest in Tanzania as a model system.

Our Closest Relative: A Killer Ape? (facing page)   Chimpanzees share 98 percent of human genes. These social, intelligent, and dextrous primates, in all probability, are closest to our common ancestors. Have humans inherited some aspects of their behavioral repertoire? And does chimp behavior serve as an analog to the dark side of human nature? Some chimps are highly predatorial, hunt cooperatively with great efficiency, and are very aggressive. However, other chimps do not kill; these chimps channel their aggression into an array of sexual behaviors.

Chimpanzees are human's closest cousins, and by some genetic estimates we share 99 percent of the same genes.

The Gombe chimps have been studied extensively. (Chapter 4 of this book described their predatory hunting behaviors in some detail). Further studies show that free-ranging chimpanzees also systematically brutalize and kill former troop members who have established a neighboring community. In these conflicts, patriarchal-male-bonded groups form coalitions to raid their neighbors, killing other males, raping and battering the females, and killing and cannibalizing their infants. Wrangham and Peterson consider the Gombe chimps', male-driven, lethal intergroup interactions as a model that reveals the ancestral origins of human brutality and warfare. Such conclusions are nothing new. Raymond Dart, whose archaeological studies revealed australopithicus's role in the evolutionary chronology of human ascent, emphasized human predatory behavior as "the mark of Cain" and, in doing so, connected hunting and carnivorousness with assault and war. Later, Robert Ardrey, in his best seller, *African Genesis,* equated human carnivorousness with the dark side of human nature, referring to humans as "killer apes."

Is there any physiological basis for the theory that male hormones alter brain function and behavior? It has been well established that testosterone primes the brain's sexual appetite centers. The operational word here is *primes,* meaning that it allows those centers to more readily respond to sensory signals that produce increased sex drive. The hormone by itself does not produce sex drive, even in high concentrations. And the priming action of testosterone occurs not only in males by also in females.

The female's sexual appetite centers are also activated by testosterone, even though women produce only one-tenth of the testosterone that men do. Could it be that testosterone has a similar priming effect on those so-called rage centers in the hypothalamus of the brain that were so elegantly mapped out by the Swiss Nobel Prize winner W. R. Hess?

We know that both the amygdala and hippocampus, structures that are known to play a role in assertive and aggressive behaviors, have the same enzyme, aromatase, that processes testosterone in the sexual appetite centers. If the male hormone simply primes the brain's aggressive machinery to respond to environmental signals and previous learning, then culture must be the primary determinant of violent aggression.

Other studies of a different species of chimpanzees, *Pan paniscus,* the pygmy chimps or bonobos, show that they exhibit a spectrum of behaviors very different from those of the Gombe chimpanzees. Although bonobos, both male and female, sometimes exhibit aggressive intents by branch dragging, stomping, and scraping, these tantrums seldom develop into an assault. It seems that the bonobos have developed a behavioral pacifier; sex play. In 1871, Charles Darwin, in his *Descent of Men and Selection in Relation to Sex,* concluded that the psyche (nervous system) regulates most body functions and behaviors including courage, perseverance, construction of strategies, and weapons of all kinds. We know from the work of Heinrich Kluver and Paul C. Bucy and others that damage to the hippocampus and amygdala causes an array of bizarre behaviors: dominant animals become submissive, monkeys lose their fear of snakes, aggressiveness is diminished, and the spectrum of sexual behaviors is blurred. Thus, it is apparent that dominance, aggression, and sex behaviors relate to the same parts of the brain.

Bonobos have apparently developed a sexual strategy that minimizes aggression and promotes bonding. Young, juvenile, and adult bonobos, of both sexes, exhibit a pervasive array of blatantly sexual acts such as frontal, that is, face to face and dorsal-ventral mock copulatory behaviors, as well as rump rubbing of genitalia, accompanied by erotic vocalizations to relax tensions and reduce competition. The behaviors are both heterosexual and homosexual; and these behaviors serve as a form of greeting, as gestures of appeasement, as a means of gaining acceptance into a group, and as a way of reassuring other members of the group that there is no hostile intent. The end result is that bonobos, unlike their cousins, the Zombe chimps, form gregarious, pacific groups with a measure of gentleness and tolerance. Trying to relate our behaviors to the sexual behaviors of other primates that diverged from a common ancestor millions of years ago is a perilous undertaking. But, given the widespread human penchant for polygamy and polyandry, it would seem that the highly publicized sexual adventures of the dominant power-seeking males that occupy the upper levels of the government of the United States (and elsewhere) should come as no surprise!

It is possible then, that evolution provided us with two endowments, the one nurturing our ability to live communally and cooperatively in peaceful coexistence with members of our own species and the other impelling us to have lives in which the dark side, with violent aggression and dominance seeking by members of a community, manifests itself.

We modern humans, descendents of a violent common primate ancestor, have incorporated into our brains the "software" for violence. However, the higher centers of our brains exert strong regulatory control over those more primitive programs lodged in the deeper recesses of the limbic system. It is these higher centers that allow both chimpanzees and humans to think and learn and that usually inhibit their assaultive behaviors.

Returning to chimpanzees as models, they have demonstrated the capacity to think, analyze, improvise, imitate, cooperate, and once having learned, retain that learning as a memory that can be recalled on demand. For example, a group of chimpanzees in Guinea was observed to perform an extraordinary feat of conscious thinking that was previously unimagined. The chimps' home range was around an enormous fig tree that was too wide for them to climb, and chimps love figs. Nearby was a climbable kopak tree, but its branches were just out of reach of the fig's. The dominant males of the group kept climbing the kopak, seeking solutions. Grappling with the problem, they created tools by breaking off branches and then tried to snare the fig tree's branches. Each near-success was greeted with hoots of encouragement from the rest of the group on the ground.

Then the number-three male solved the problem. He began rapidly breaking small branches off the large branch of the kopak tree on which he was standing. The denuded branch, now lighter, sprang up, carrying him high enough to reach out and grab a branch of the fig tree and pull it down. The number two male then joined him so that their combined weight would bend the branch low enough for the whole group to clamber up the tree and "fig out," as it were. Such a capacity to learn, think, and modify behavior indicates that neural circuits can be reshaped by learning, learning that can be communicated to others. Other genes, favoring long life spans, provided for the accumulation of experiential learning, complementing the capacity to learn.

Females and Aggression

In the past, human cultures have generally been permissive of males' acting abominably toward females. But today, in some societies, this dominance has eroded, and rules of fair play have been enacted into laws that finally permit females to compete and bring into play their brain centers that regulate aggression. Such leveling of the playing fields is only a relatively recent phenomenon. Males steeped in older traditions are resentful and slow to accept female rights, and even when they do, they do so grudgingly. What is the problem? Much of it has to do with male upbringing and culture. Prior to puberty, young boys are literally free of any influences of testosterone on their brains; yet their play patterns usually are combative, and their toys weapon oriented. Even in families where toy guns and warlike toys are prohibited, the male children improvise weapons and war games. These games have rules which impose limits on behavior. Later, as boys are introduced to organized, competitive contact sports (but still before puberty), the rules are made more stringent and violations of fair play call for stiff penalties. So, in growing up, males have had imposed on their aggressive competitive drives a series of restraints similar to those seen in many animal societies, restraints that animal behaviorists call "chivalrous inhibition."

Females, too, have been raised in cultures that influenced their later behavior, and when competition was present, it was channeled into different avenues. Herein lies the problem: the new societal changes permitting women to compete with men do not recognize that female aggressive behavior *sometimes* lacks the early learning to practice chivalrous inhibition, and males don't know how to cope with female aggression. These interactions are further complicated by the fact that females and males have learned to view each other within a framework of choosing a mate. Both males and females have evolved mechanisms, strategies, and tactics to influence each other. Male tactics coevolved with female tactics, but females often set the conditions for sexual selection. Given this background of the coevolution of intricate behaviors that ensure that the species will reproduce and be preserved, these behaviors are always present at some level when males and females interact. Adding a sexual component to the historical component of male dominance certainly makes interaction in the work place even more difficult, creating an antagonistic mind-set.

If we modify the definition of assaultive aggression to include those behaviors which cause not only physical but also emotional hurt to others or physical damage to inanimate objects, we find that female patterns of aggression mirror those of men and boys. According to many recent studies, females in domestic disputes are just as likely to be assaultive as men, although often their weapons are words, not fists. Yet a detailed, large-scale study of 3,000 young women in Canada, the United States, Britain, and New Zealand reported that women were even slightly more likely to use some form of physical attack than men. (However, because men have 20 percent more muscle than women, male assaultive behavior produces more physical damage.)

At a 1998 meeting of the International Society for Research on Aggression, some investigators reported that even in childhood there are substantial similarities between female and male aggression, a finding that runs counter to current beliefs. Moreover, recently when investigators included verbal assault in the spectrum of assaultive behaviors, they found that it can do even greater long-term harm to the victim than a punch, slap, or kick. Verbal aggression is more prevalent in females than in males, and some reports suggest females persevere longer in the attack mode than males. The old concept of war between the sexes and the militancy exhibited by some of the extremists in this evidently phony conflict are often popularized by the media and add little to mutual understanding.

The Ultimate Triumph of the Left Brain

In the past century, humans have been involved in violent, massive conflicts that caused death and suffering on a scale never seen before. These conflicts brought to the fore ever more lethal weapons systems, including nuclear, chemical, and biological weapons—weapons of mass destruction. Warfare has always squandered human inventiveness by directing it into weapons technologies. The development of each new system is like opening Pandora's box, and the consequences are often catastrophic.

The problem is that new technologies are now being created at a pace that places a strain on humans' capacity to adapt and modify their behaviors. It seems that once we've let the evil genie out of the bottle, we humans try to control it. World bodies like the League of Nations and the United Nations, as well as treaties between nations, have attempted, with varying success, to impose limits on nuclear weaponry and prevent their proliferation. It seemed to be working until India and Pakistan, at great cost, launched nuclear tests, to the cheers of their populations, which considered such weaponry a symbol of status. Why do countries that lack safe water supplies and flush toilets and are ravaged by poverty waste their limited resources?

On the other hand, it certainly seems that the collective wisdom of humans' thinking brains has a truly noble side filled with caring and compassion. Natural disasters, epidemics, and starvation always bring forth international efforts to help those who are suffering. We have also, with some degree of success, initiated international efforts to prevent and limit bloody internecine conflict, and international law is now reaching out to punish those involved in genocide. At the grass roots level and at national and international levels, we support programs geared toward the preservation of endangered species. We fight atmospheric pollution and are attempting to do something about the greenhouse gases and the coolants that are destroying the ozone layer. Organizations and nations are active in attempting to ensure human rights and at the same time do something to regulate the explosive growth of the human population.

Notwithstanding their origin, the technologies spawned by weapons development are having a massive impact on improving the quality of human life. If we look to the positive accomplishments of humans, we must feel optimistic about the future. There will still be violent people; variability, after all, is a characteristic of living things. But the frequency of violent behaviors seems to be decreasing. We are not descendants of the killer ape. We are, after all is said and done, *Homo sapiens*—the flawed wise human animal.

Index